Test Yourself

Electric Circuits

Mehdi Anwar, Ph.D.
Electrical and Systems Engineering
University of Connecticut
Storrs, CT

Contributing Editors

Thomas Hall, M.S.
Northwestern State University of Louisiana
Natchitoches, LA

Nadipuram R. Prasad, Ph.D.
Electrical and Computer Engineering
New Mexico State University
Las Cruces, NM

Leane Roffey, Ph.D.
NeuroMagnetic Systems
San Antonio, TX

NTC LEARNINGWORKS
NTC/Contemporary Publishing Group

Library of Congress Cataloging-in-Publication Data

Anwar, Mehdi.
Electric circuits / Mehdi Anwar.
p. cm. — (Test yourself)
ISBN 0-8442-2354-9
1. Electric circuits—Examinations, questions, etc. 2. Electric circuit analysis—Examinations, questions, etc. I. Title.
II. Series: Test yourself (Lincolnwood, Ill.)
TK454.A59 1997 97-25148
CIP

A *Test Yourself Books, Inc.* **Project**

Published by NTC LearningWorks
A division of NTC/Contemporary Publishing Group, Inc.
4255 West Touhy Avenue, Lincolnwood (Chicago), Illinois 60646-1975 U.S.A.

Printed in the United States of America
International Standard Book Number: 0-8442-2354-9
18 17 16 15 14 13 12 11 10 9 8 7 6 5 4 3 2 1

Contents

Preface

This "Test Yourself" book, *Electric Circuits*, is written for students who are taking the first course in linear circuit analysis. The problems are standard, and the students may face problems of this nature on any one of their examinations.

In chapter 1, simple resistive circuits under dc excitation are solved and the concept of series/parallel circuit elements is introduced. That is followed by network theorems in dc circuits in chapter 2. The methods of nodal and mesh analysis are used to solve rather complicated dc circuits. These methods are applied in solving circuits using Thevenin's/Norton's and superposition principles. Source transformations are also introduced to help simplify circuits. Power dissipation in resistors and absorption/dissipation in independent/dependent sources are discussed.

Circuit transients are extremely important during switching. An exact solution of circuit transient requires the ability to solve first- and second-order linear differential equations. Transients in RC, RL, and RLC circuits are discussed in chapter 3.

In chapter 4, network theorems introduced in chapter 2 are extended to ac circuits. The calculation using the network theorems gives the steady-state response. In general, the total response in any ac circuits is a combination of the transient and the steady-state responses. Circuits containing operational amplifiers are introduced next. Both ac and dc circuits are considered.

Chapter 6 introduces the method of phasors in solving ac circuits. This particular method is only applicable under sinusoidal excitation and always gives the steady-state response. Phase in ac circuits results in the concept of real *and* reactive power (in contrast to only real power in dc circuits). Real, reactive, and apparent powers are represented in a power triangle. The goal of an electrical engineer is to maximize the real power delivered to a load. This is done by connecting capacitors at critical points, thereby improving the power factor angle. This important technique is presented in chapter 7.

Chapter 8 provides some typical Spice simulations and their results. Important problems using operational amplifiers are solved. Codes to study switching transients are provided.

Mehdi Anwar, Ph.D.

How to Use This Book

This "Test Yourself" book is part of a unique series designed to help you improve your test scores on almost any type of examination you will face. Too often, you will study for a test—quiz, midterm, or final—and come away with a score that is lower than anticipated. Why? Because there is no way for you to really know how much you understand a topic until you've taken a test. The *purpose* of the test, after all, is to test your complete understanding of the material.

The "Test Yourself" series offers you a way to improve your scores and to actually test your knowledge at the time you use this book. Consider each chapter a diagnostic pretest in a specific topic. Answer the questions, check your answers, and then give yourself a grade. Then, and only then, will you know where your strengths and, more important, weaknesses are. Once these areas are identified, you can strategically focus your study on those topics that need additional work.

Each book in this series presents a specific subject in an organized manner, and although each "Test Yourself" chapter may not correspond to exactly the same chapter in your textbook, you should have little difficulty in locating the specific topic you are studying. Written by educators in the field, each book is designed to correspond, as much as possible, to the leading textbooks. This means that you can feel confident in using this book, and that regardless of your textbook, professor, or school, you will be much better prepared for anything you will encounter on your test.

Each chapter has four parts:

Brief Yourself. All chapters contain a brief overview of the topic that is intended to give you a more thorough understanding of the material with which you need to be familiar. Sometimes this information is presented at the beginning of the chapter, and sometimes it flows throughout the chapter, to review your understanding of various *units* within the chapter.

Test Yourself. Each chapter covers a specific topic corresponding to one that you will find in your textbook. Answer the questions, either on a separate page or directly in the book, if there is room.

Check Yourself. Check your answers. Every question is fully answered and explained. These answers will be the key to your increased understanding. If you answered the question incorrectly, read the explanations to *learn* and *understand* the material. You will note that at the end of every answer you will be referred to a specific subtopic within that chapter, so you can focus your studying and prepare more efficiently.

Grade Yourself. At the end of each chapter is a self-diagnostic key. By indicating on this form the numbers of those questions you answered incorrectly, you will have a clear picture of your weak areas.

There are no secrets to test success. Only good preparation can guarantee higher grades. By utilizing this "Test Yourself" book, you will have a better chance of improving your scores and understanding the subject more fully.

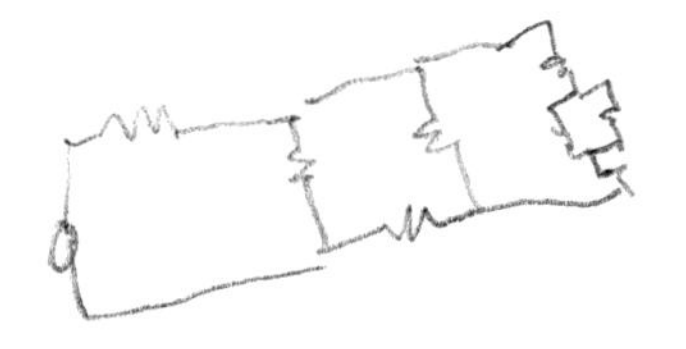

Simple Resistive Circuits

Brief Yourself

Simple resistive circuits under dc excitation are introduced in this chapter. The chapter introduces Kirchoff voltage and current laws (KVL and KCL) to solve problems involving series and parallel resistors. The concept of dependent sources is introduced in this chapter. $\Delta - Y$ transformation is used in solving problems. Power supplied by dependent/independent sources and that dissipated by resistors are calculated.

Test Yourself

1.

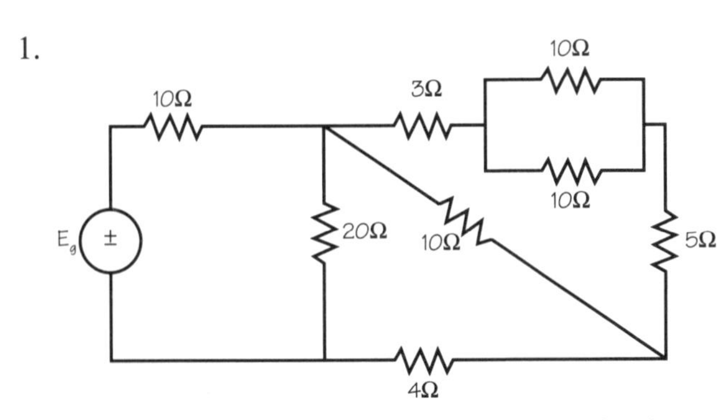

Determine the equivalent resistance seen by the source.

2.

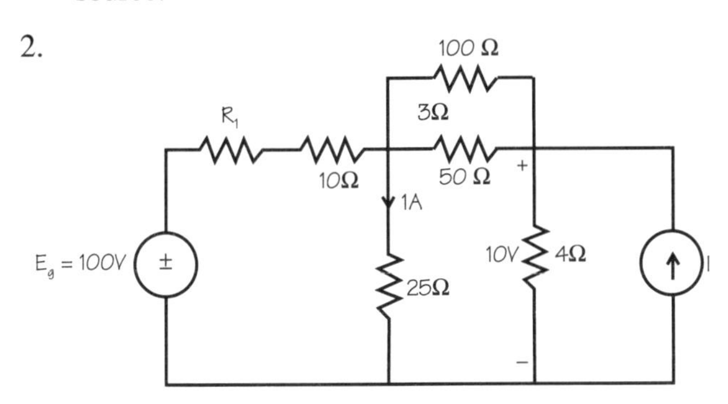

For the circuit shown, determine the voltage across the 100Ω resistor. Also determine the magnitude of R_1 and I.

3.

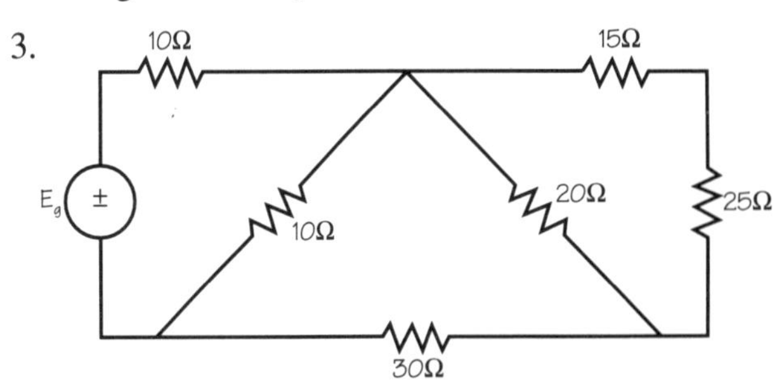

Determine the equivalent resistance seen by the source.

4.

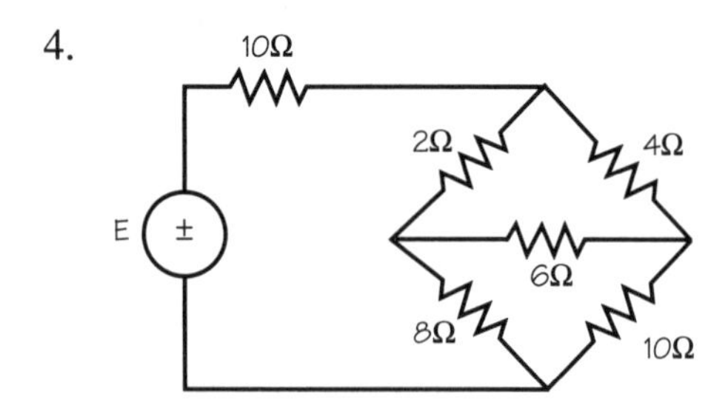

For the bridge circuit shown, determine the equivalent resistance seen by the voltage source.

5. In problem 1.2, calculate the power supplied by the sources.

6.

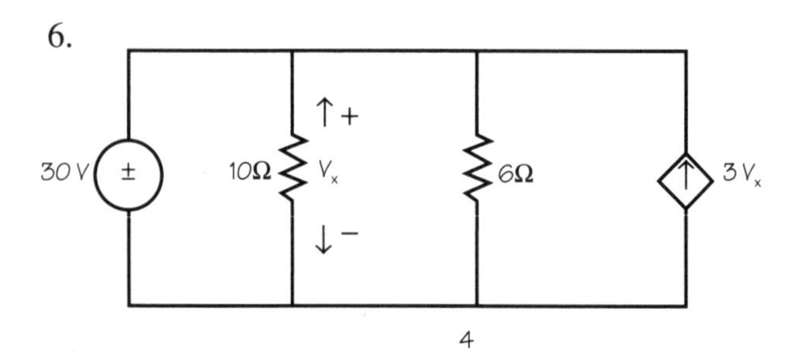

Determine the power supplied by the independent source. Is there any power supplied by the dependent source?

7.

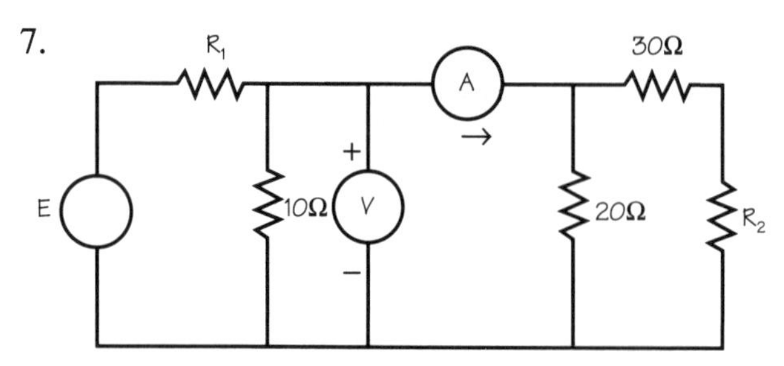

The voltmeter and the ammeters read 20V and 1A, respectively. Calculate R_1 and R_2, given the power supplied by the source equals 600W. Assume ideal meters.

8.

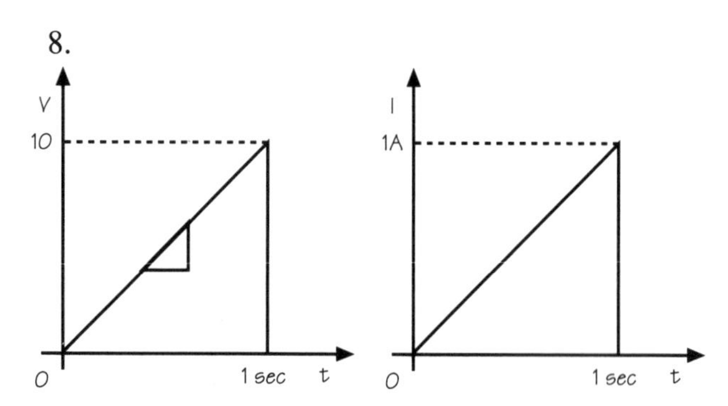

For a series-resistive circuit, the input voltage and the resulting current shapes are shown. Calculate the resistance.

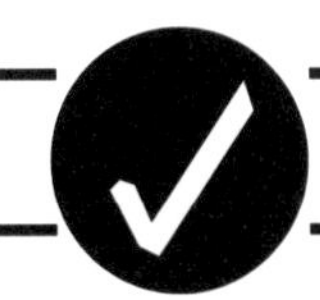

Check Yourself

1.

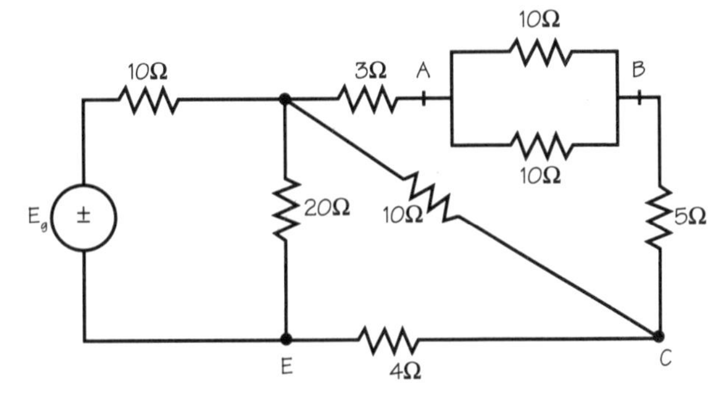

The two 10Ω resistances (between points A and B) are in parallel.

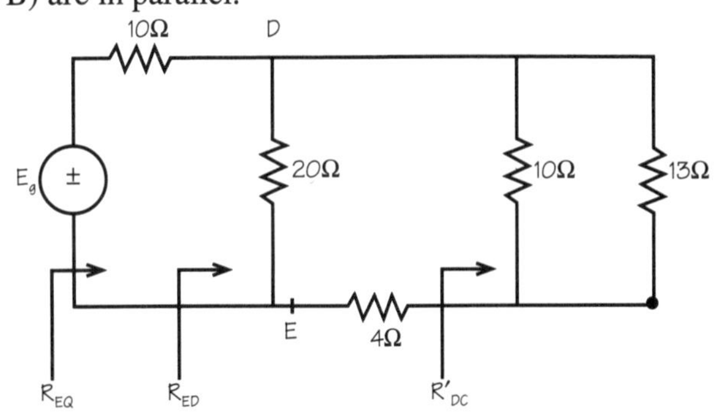

$$R_{AB} = \frac{10 * 10}{10 + 10} = 5\ \Omega$$

$$R_{AC} = R_{AB} + R_{BC} = 5 + 5 = 10\Omega$$

$$R_{DC} = R_{DA} + R_{AC} = 3 + 10 = 13\Omega$$

$$R'_{DC} = 13 \| 10 = 5.652\Omega$$

$$R_{ED} = 20\|(R'_{DC} + 4) = \frac{20 * 9.652}{20 + 9.652}$$

$$= 6.51\Omega$$

The equivalent resistance seen by Eg, R_{EQ} is:

$$R_{EQ} = 10 + R_{ED} = 16.51\Omega$$

(Series-parallel)

2.

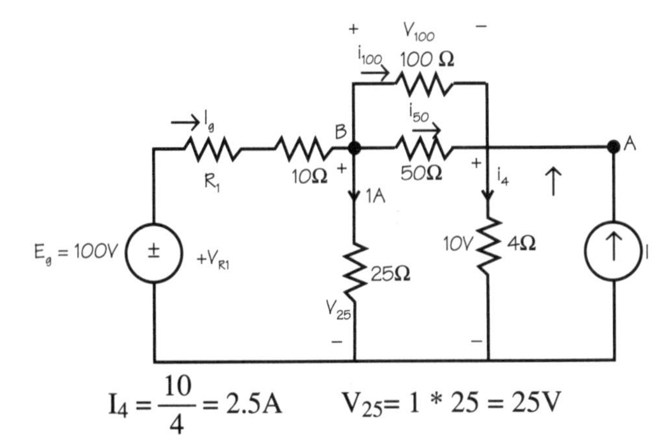

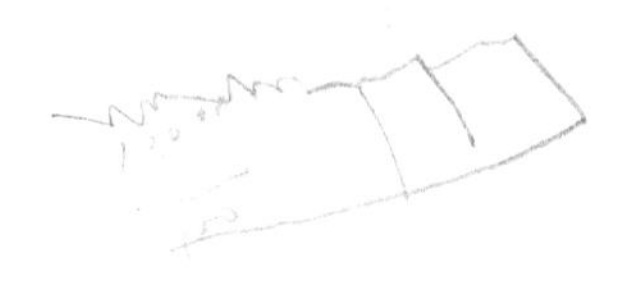

$$I_4 = \frac{10}{4} = 2.5A \qquad V_{25} = 1 * 25 = 25V$$

Applying KVL:

$$V_{25} - V_{100} - 10 = 0$$

$$V_{100} = V_{25} - 10 = 25 - 10 = 15V$$

$$I_{50} = \frac{V_{100}}{50} = \frac{15}{50} = 0.3A$$

$$I_{100} = \frac{15}{100} = 0.15A$$

Applying KCL at B:

$$Ig = \frac{V_{25}}{25} + 1_{100} + I_{50}$$
$$= 1 + 0.15 + 0.3 = 1.45A$$
$$V_{RI} = 100 - V_{25} - 10 * Ig = 100 - 25 - 14.5 = 60.5$$
$$R_I = \frac{60.5}{1.45} = 41.72\Omega$$

Applying KCL at A:

$$I = I_4 - I_{50} - I_{100}$$
$$= 2.5 - 0.3 - 0.15 = 2.05A$$

$$V_{100} = 15V$$
$$R_I = 41.72\Omega$$
$$I = 2.05\Omega$$

(Branch circuit)

3.

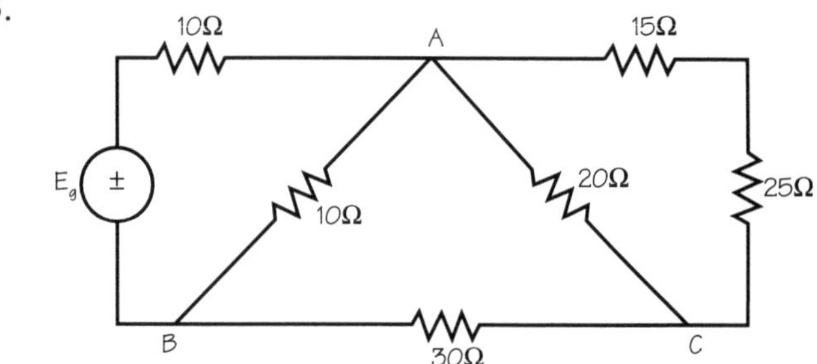

$$R_{AC} = 20||(15 + 25) = \frac{20 * 40}{20 + 40} = 13.33\Omega$$

The circuit reduces to:

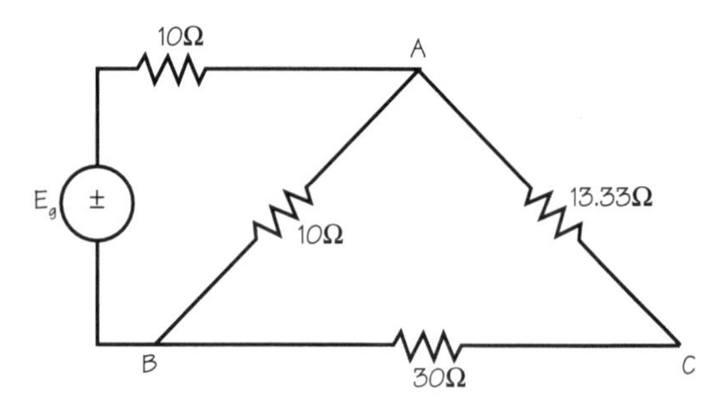

The delta-connected ABC can be transformed to a Y to simplify the circuit.

$$R_1 = \frac{10*13.33}{10 + 13.33 + 30} = 2.5\Omega$$
$$R_2 = \frac{10 * 30}{10 + 13.33 + 30} = 5.63\Omega$$
$$R_3 = \frac{30 * 13.33}{10 + 13.33 + 30} = 7.5\Omega$$

The circuit can now be redrawn as:

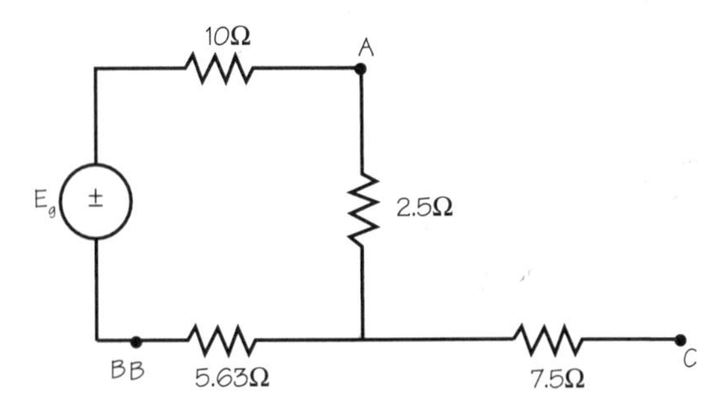

$$R_{eq} = 10 + 2.5 + 5.63 = 18.13\Omega$$

A simpler solution can be obtained as:

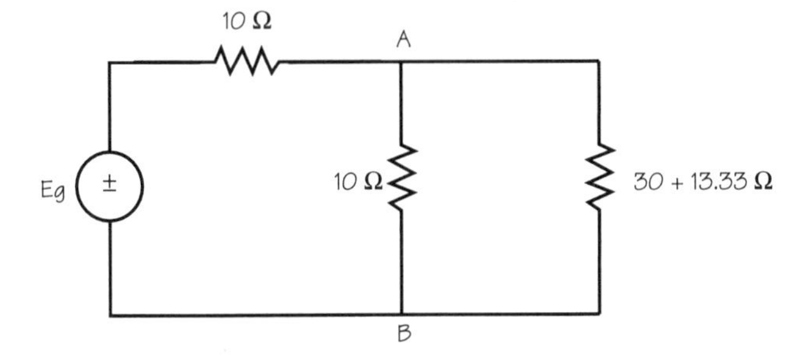

$$R_{eq} = 10 + 10 \| 43.33 = 18.13\ \Omega$$

(Δ – Y)

4.

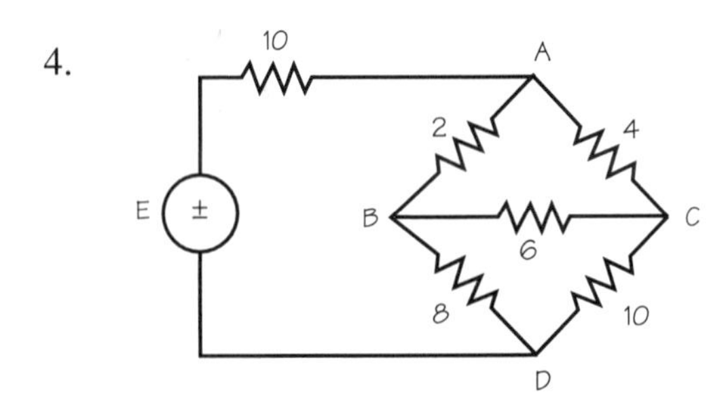

Transforming the ABC $\triangle$ to a Y, we obtain:

$$R_A = \frac{2 * 4}{2 + 4 + 6} = 0.667\Omega$$

$$R_B = \frac{2 * 6}{2 + 4 + 6} = 1\Omega$$

$$R_C = \frac{6 * 4}{2 + 4 + 6} = 2\Omega$$

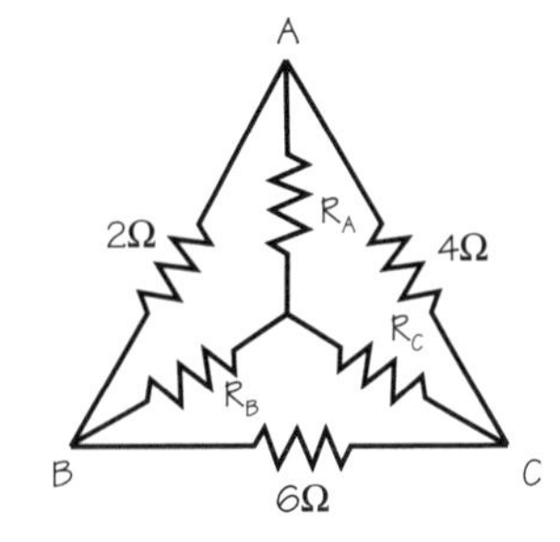

The circuit can now be redrawn as:

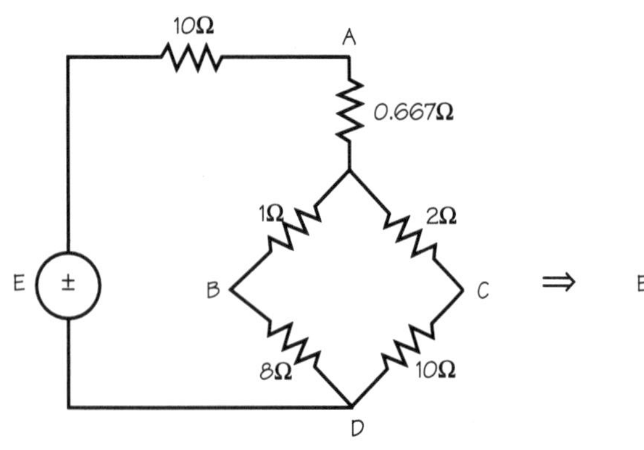

$$R_{eq} = 10 + 0.667 + 9||12 = 15.809\Omega$$

(Δ – Y)

5. From problem 1.2

   Power supplied by the sources:

   $P_{Eg} = 100 * 1.45 = (+)145W$
   $P_I = 2.05 * 10 = (+)20.5W$

   (Power calculation)

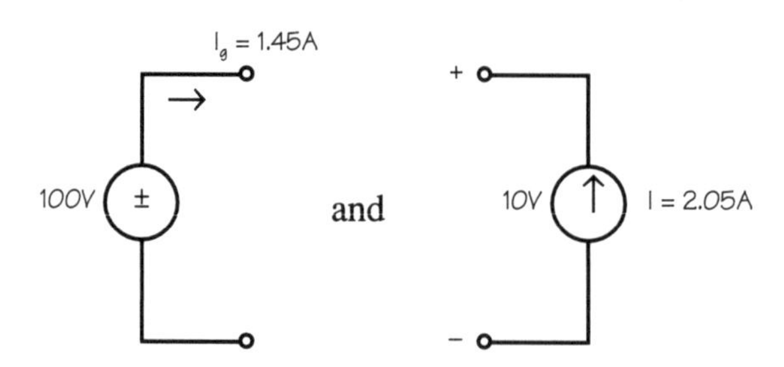

6.

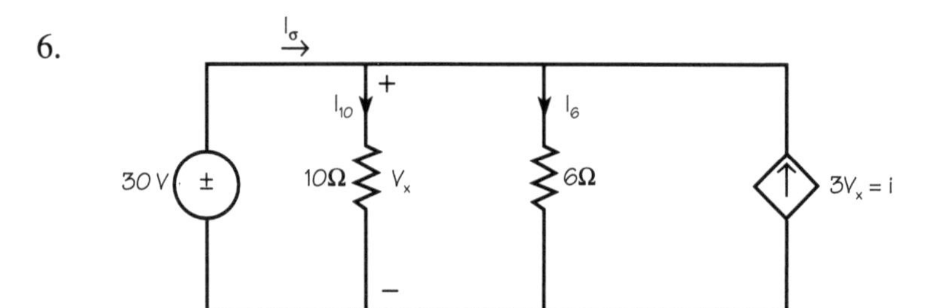

From the circuit, $V_X = 30V \quad i = 3V_X = 90A$

$$I_{10} = \frac{V_X}{10} = 3A$$

$$I_6 = \frac{30}{6} = 5A$$

$$I_\sigma = I_{10} + I_6 - i = 3 + 5 - 90 = -82A$$

Power supplied by the sources:

$P_{30} = 30 * I_\sigma = -30 * 82 = -2460W$ (absorbing power)
$P_i = 30 * 90 = +2700W$

(Power calculation)

7.

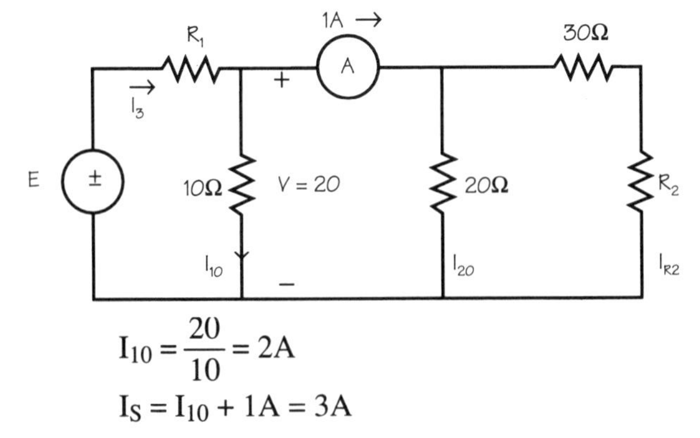

$$I_{10} = \frac{20}{10} = 2A$$
$$I_S = I_{10} + 1A = 3A$$

Power supplied by the source:

$$P_E = 600W = E * I_S \Rightarrow E = \frac{600W}{3A} = 200V$$

Therefore, $R_1 = \frac{200 - 20}{3} = \frac{180}{3} = 60\Omega$

$$I_{20} = \frac{20}{20} = 1A, I_{20} + I_{R2} = 1A$$
$$I_{R2} = 1 - I_{20} = 0A$$

R_2 is an open circuit.

(DC circuits)

8. $R_2 = \frac{V}{I} = \frac{10A}{1A} = 10\Omega$

(DC circuits)

Grade Yourself

Circle the numbers of the questions you missed, then fill in the total incorrect for each topic. If you answered more than three questions incorrectly, you need to focus on that topic. (If a topic has less than three questions and you had at least one wrong, we suggest you study that topic also. Read your textbook, a review book, or ask your teacher for help.)

Subject: Simple Resistive Circuits

Topic	Question Numbers	Number Incorrect
Series-parallel	1	
Branch circuit	2	
$\Delta - Y$	3, 4	
Power calculation	5, 6	
DC circuits	7, 8	

Network Theorems

2

Brief Yourself

This chapter uses the different network theorems to solve simple circuits. The methods of mesh currents (MC) and node voltages (NV) are basic methods and are based upon the technique of branch currents (BC). Thevenin's/Norton's equivalent circuit representation may be made tractable by using any of the basic methods MC/NV/BC etc. Thevenin's/Norton's methods are more realistic in the sense that, in real measurements, we either measure voltage between two points (first by disconnecting the non-required portion of the circuit) that can be related to the Thevenin's voltage or measure resistance as it follows by first turning off live sources. The method of superposition is a simple application of the superposition theorem valid for linear systems. Source transformation provides a way to simplify complicated circuits.

Test Yourself

1. Using the method of mesh currents, calculate I_0.

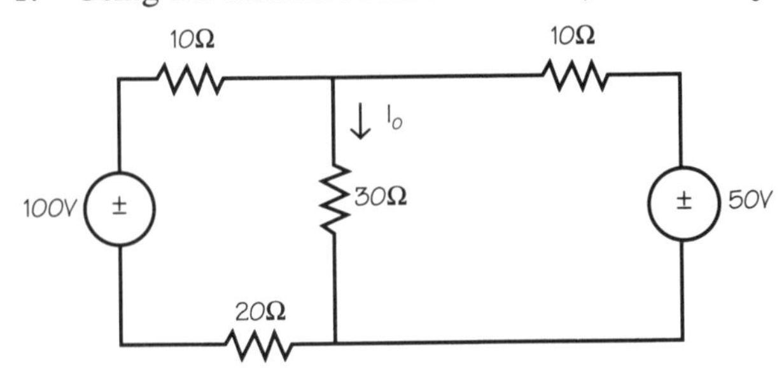

2. Using the method of mesh currents, calculate I_2.

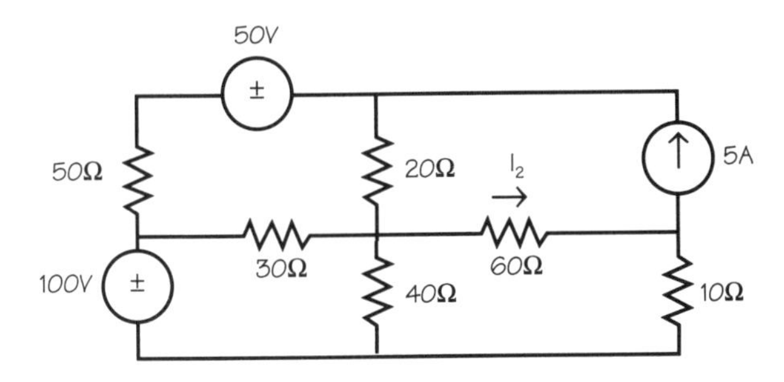

3. Calculate the power supplied by the 100V source (use mesh currents).

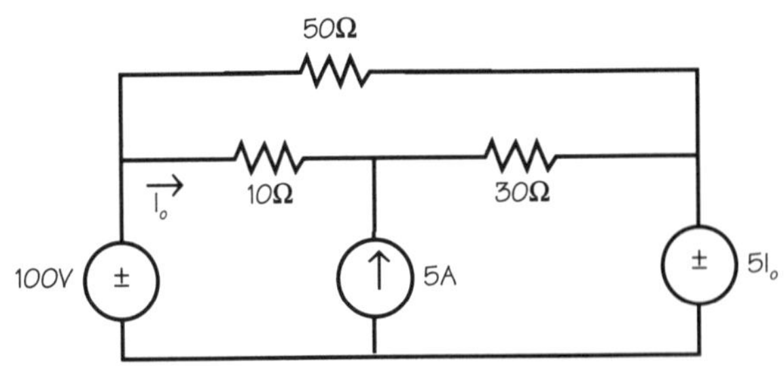

4. Calculate V_0 in terms of V_s using mesh analysis.

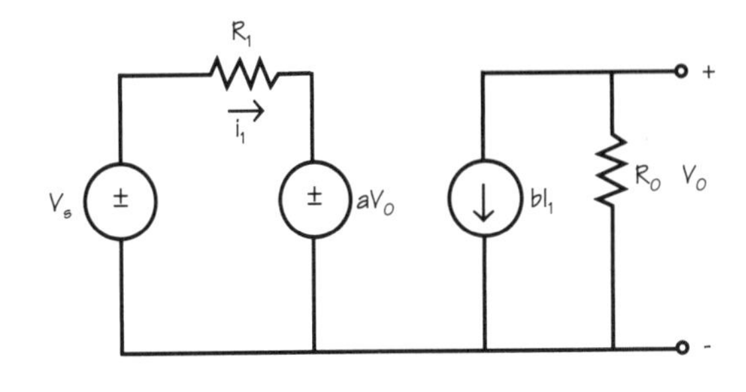

5. Calculate the power supplied by the current source.

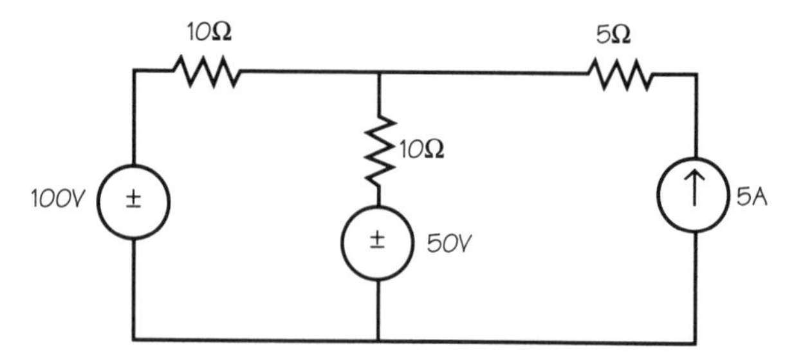

6. For the bridge circuit shown, calculate the node voltages.

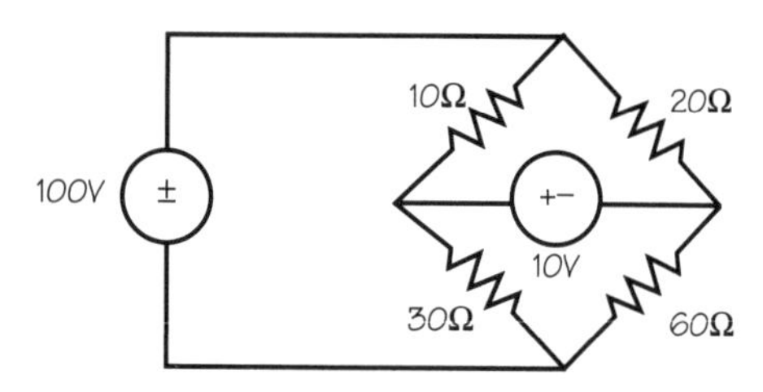

7. Using the method of node voltages, solve for all the node voltages.

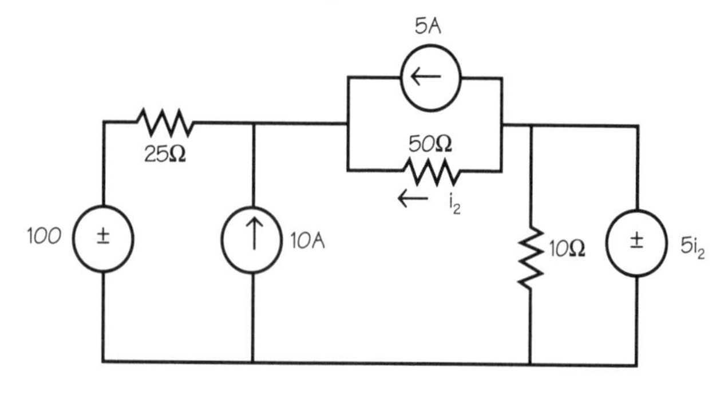

8. Calculate I by replacing the circuit between *a* and *b* by its Thevenin's equivalent.

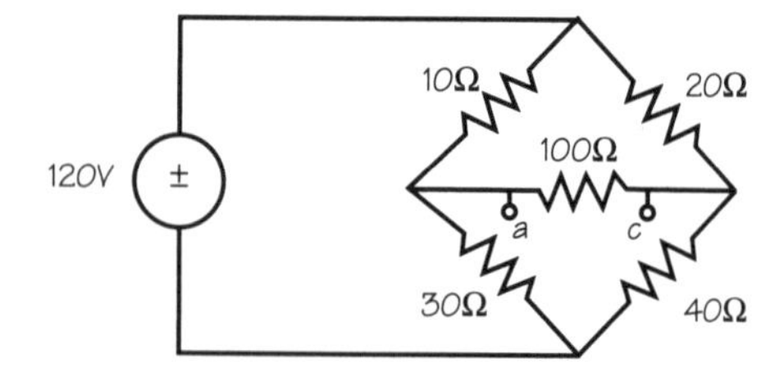

9. Replace the circuit to the left of *a-b* by its Thevenin's equivalent.

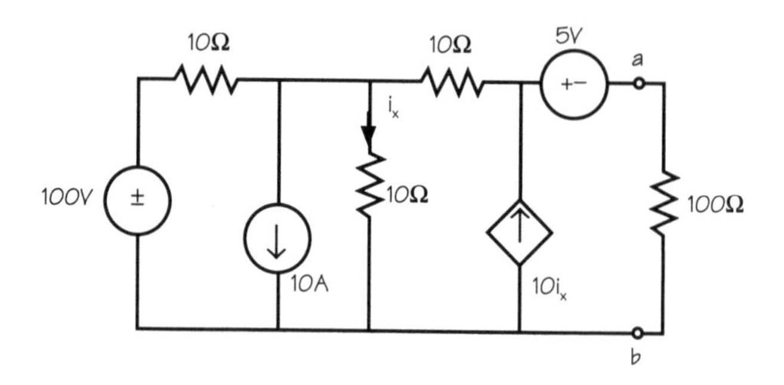

10. Replace the circuit shown below by its Thevenin's equivalent to the left of *a-b* (the circuit shown is the small signal representation of bipolar junction transistors).

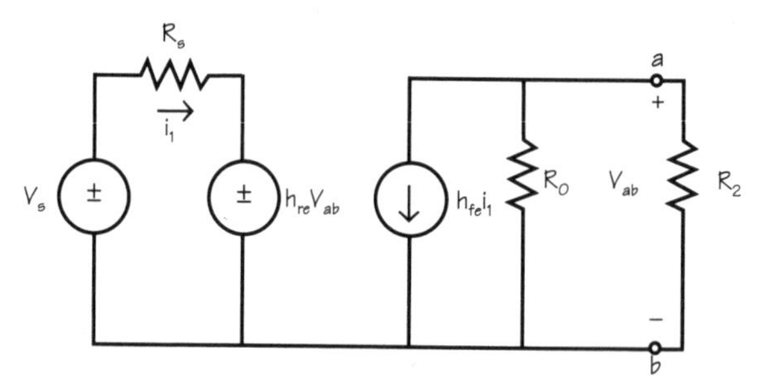

11. For the circuit shown, determine its Thevenin's equivalent, looking into terminals *a* and *b*.

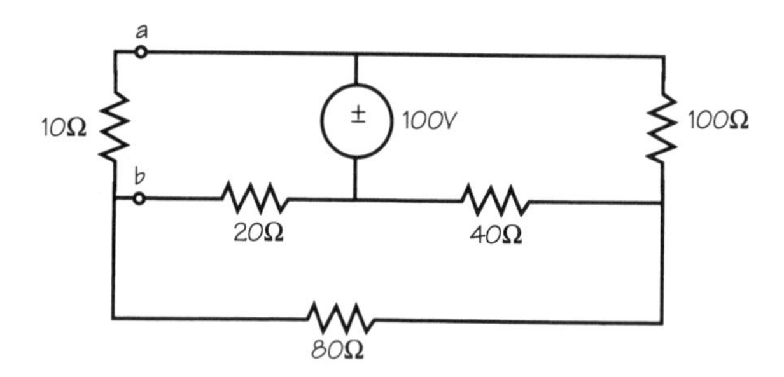

12. Calculate current I_{ab}, by replacing the rest of the circuit by a Norton's equivalent.

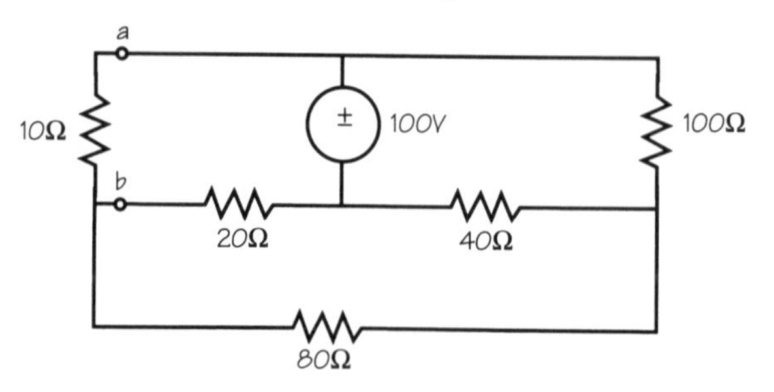

13. For the circuit shown in problem 2.12, calculate the load resistance (between terminals *a* and *b*) for maximum power transfer.

14. Find the current I using superposition theorem.

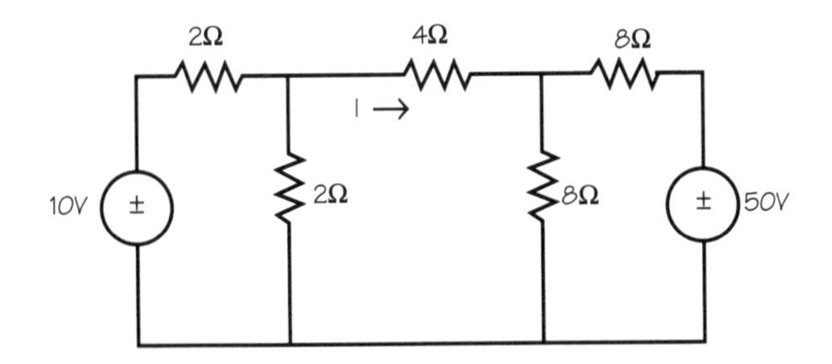

15. Using the method of superposition, calculate I.

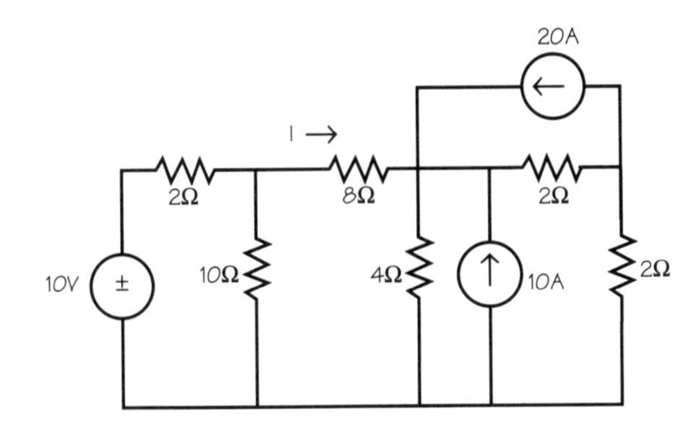

16. Use superposition to calculate the voltage V.

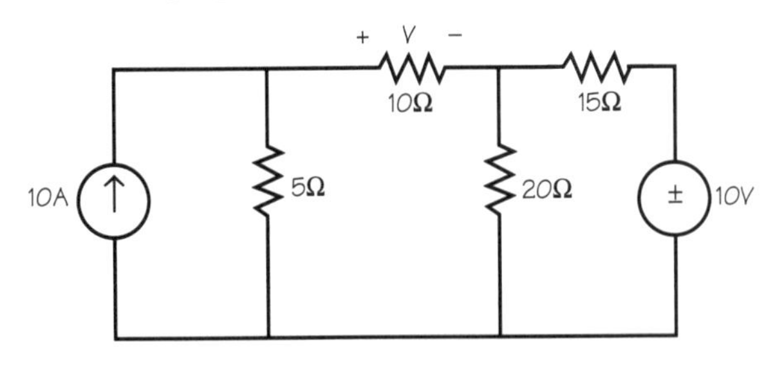

17. Use source transformation to solve for V in the circuit shown in problem 2.16.

Check Yourself

1. Mesh current equation:

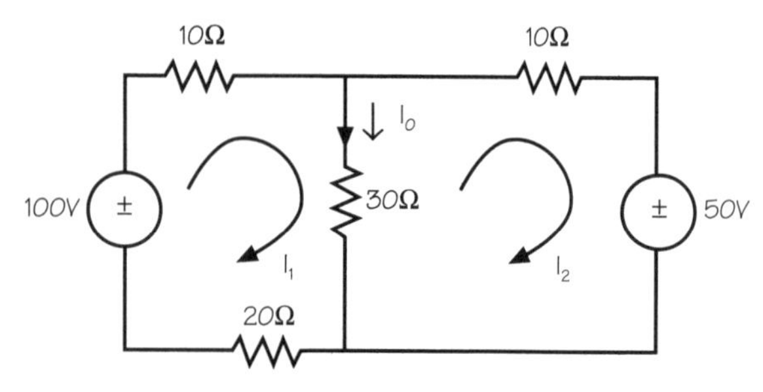

$$100 = 60\ I_1 - 30\ I_2$$
$$-50 = -30\ I_1 + 40\ I_2$$

Using Cramer's rule =: $I_1 = \dfrac{\begin{vmatrix} 100 & -30 \\ -50 & 40 \end{vmatrix}}{\begin{vmatrix} 60 & -30 \\ -30 & 40 \end{vmatrix}} = 5/3\ A$; $I_2 = 0A$; $I_0 = I_1 - I_2 = 5/3A$.

(Mesh current)

2.

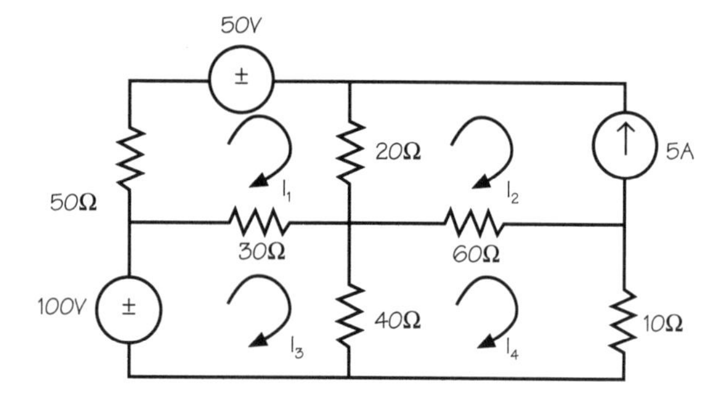

Mesh current $I_2 = -5A$.

$$-50 = 100I_1 - 20I_2 - 30I_3$$
$$100 = -30I_1 + 70I_3 - 40I_4$$
$$0 = -60I_2 - 40I_3 + 100I_4$$

In matrix notation: $\begin{pmatrix} 10 & -3 & 0 \\ -3 & 7 & -4 \\ 0 & -4 & 10 \end{pmatrix}\begin{pmatrix} I_1 \\ I_3 \\ I_4 \end{pmatrix} = \begin{pmatrix} -15 \\ 10 \\ -30 \end{pmatrix}$

Using Cramer's rule: $I_4 = \dfrac{\begin{vmatrix} 10 & -3 & -15 \\ -3 & 7 & 10 \\ 0 & -4 & -30 \end{vmatrix}}{\begin{vmatrix} 10 & -3 & 0 \\ -3 & 7 & -4 \\ 0 & -4 & 10 \end{vmatrix}} = \dfrac{10(-210 + 40) + 3(+90) + 15(12)}{10(70 - 16) + 3\ (-30)} = -2.78A$

$I_x = I_4 - I_2 = 2.22A$. **(Mesh current)**

3.

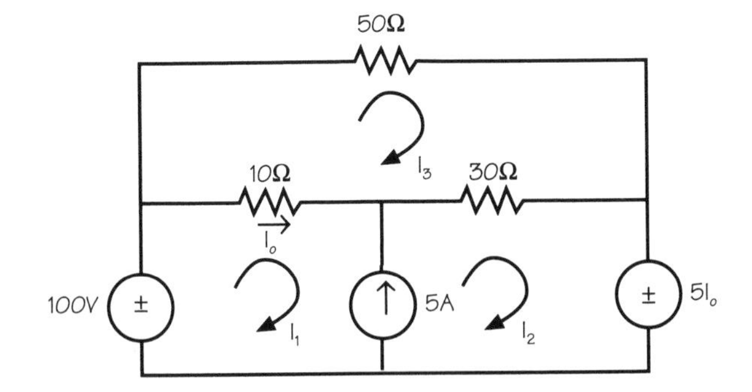

Identifying mesh defined by currents I_1 and I_2 as supernodes, we can write:

$$100 - 5\ I_0 = 10\ I_1 + 30\ I_2 - 10\ I_3 - 30\ I_3$$

reduces to $5 = 6\ I_1 - 9\ I_3$. (using $I_2 = I_1 - 5$ and $I_0 = I_1 - I_3$)
The other independent equation is: $15 = -4\ I_1 + 9\ I_3$. Giving $I_1 = 7A$.
Power supplied by the voltage source P = 100V * 10A = 700W.

(Mesh current)

4. $V_s = aV_0 + I_1 * R1 \Rightarrow I_1 = (Vs - aV_0)/R_1$

At the output, $V_0 = -R_0 * b * I_1 \Rightarrow V_0 = -R_0 * b * (V_s - aV_0)/R_1$

$$V_0/V_s = -R_1 * R_0 * b/(R_1 - a * b * R_0)$$

(Mesh current)

5.

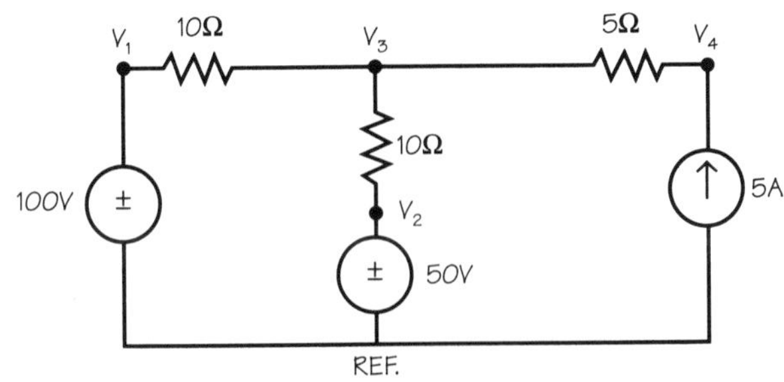

The node voltages $V_1 = 100V$ and $V_2 = 50V$ are known. Now, writing the node voltage equations at nodes V_3 and V_4, respectively, we obtain:

$$V_3(1/10 + 1/10 + 1/5) - V_1(1/10) - V_2(1/10) - V_4(1/5) = 0, \text{ that reduces to}$$
$$\Rightarrow +V_3(1/5) - V_4(1/5) = 5. \text{ At node } V_4\text{: } V_4(1/5) - V_3(1/5) = 5.$$

In matrix notation: . $\begin{pmatrix} 4 & -2 \\ -1 & 1 \end{pmatrix}\begin{pmatrix} V_3 \\ V_4 \end{pmatrix} = \begin{pmatrix} 150 \\ 25 \end{pmatrix}$. Implying $V_4 = 125V$.

Power supplied by the current source P = 125 * 5 = 625W.

(Nodal analysis)

6. 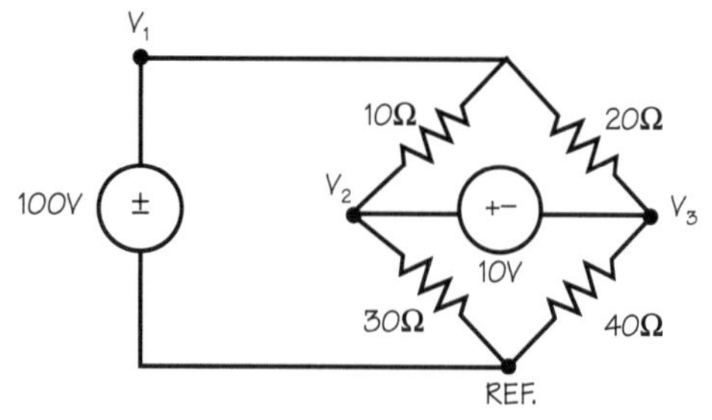

V_1 = 100V and $V_2 = V_3$ + 10V. Identifying nodes V_2 and V_3 as supernodes:

$V_2(1/10 + 1/30) + V_3(1/20 + 1/40) - 100(1/10 + 1/20) = 0$. Upon simplification V_3 = 65.6V.

V_2 = 75.6V. **(Nodal analysis)**

7. 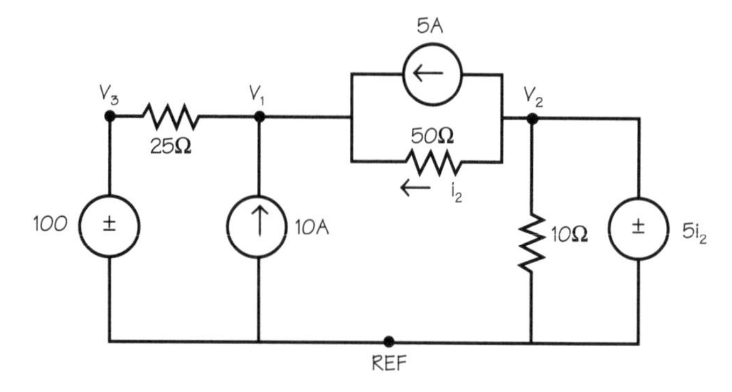

Node voltage V_3 = 100V. The current $I_2 = \dfrac{(V_2 - V_1)}{50}$ and $V_2 = 5\ I_2$, implying $9\ V_2 = -V_1$.
The node voltage equation for V_1:

$V_1(1/25 + 1/50) - 15 - V_2/50 - 4 = 0$. Using the condition $9\ V_2 = -V_1$, one obtains
$V_2 = -33.9V + V_1 = 305.4V$ **(Nodal analysis)**

8. Thevenin's voltage (open-circuit voltage)
The calculation of the open-circuit voltage between *a* and *b* is calculated by using the following circuit, obtained by replacing the resistor between *a* and *b* by an open circuit.

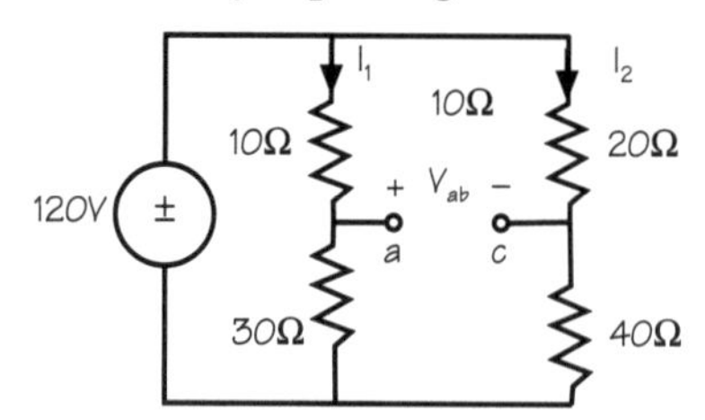

$I_1 = 120/40 = 3A$; $I_2 = 120/60 = 2A$. Applying KVL: $V_{ab} + 10 * 3 - 20 * 2 = 0 \Rightarrow V_{ab} = V_{th} = 10V$.

The calculation of Thevenin's resistance R_{TH} follows the circuit shown. The circuit shows the practical way of measuring resistance. The measurement of resistance in practical circuits proceeds by first deenergizing the circuit, which is achieved by replacing the independent voltage source by its internal resistance (that happens to be a short).

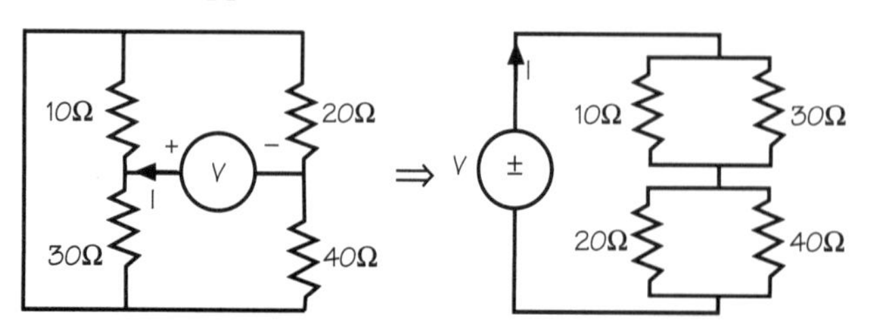

$R_{th} = 30||10 + 20||40 = 125/6\Omega$. The resulting Thevenin's circuit is: I = 10/120.8 = 82.8mA

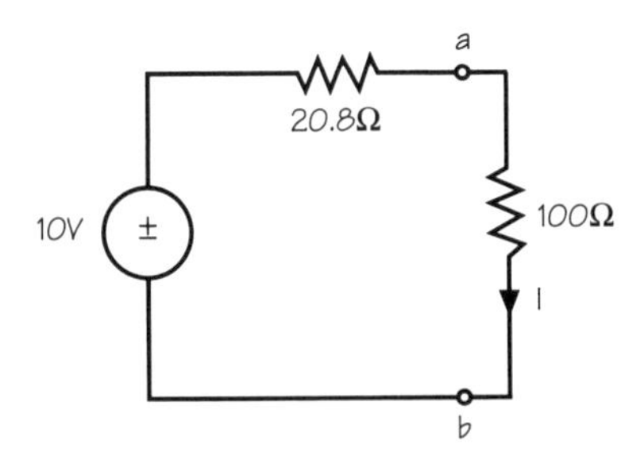

(Thevenin's theorem)

9. The open-circuit voltage Vab (or the Thevenin's voltage) is calculated by using the method of node voltages:

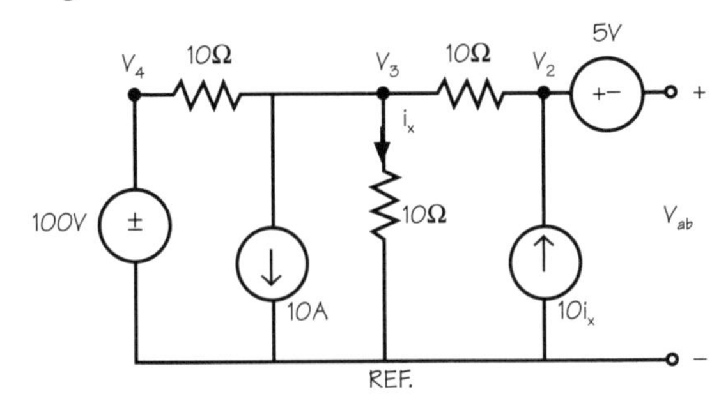

$-V_3[1/10] + V_2[1/10] + 10ix = 0$; with $ix = V_3/10$(from the circuit) $\Rightarrow V_2 = -9V_3$.
$V_3[1/10 + 1/10 + 1/10] - 100/10 - V_2[1/10] = 0$. Using the information $V_2 = -9V3$, one obtains $V_3 = 100/12V$. $V_2 = -75V$ and $V_{ab} = V_2 - 5 = -75 - 5 = -80V$ ($=V_{th}$).

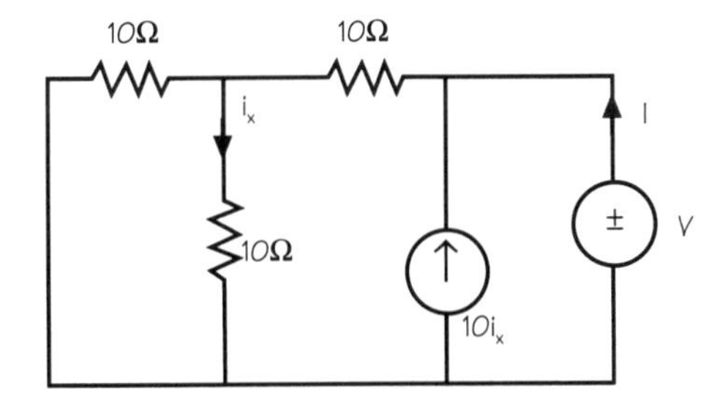

R_{th} calculaton is based upon the following circuit:

$i_x = \frac{V}{10}$. Applying KCL at A

$$I = -10i_x + \frac{V - V_I}{10} + \frac{V}{10} - \frac{V_I}{5} \qquad \text{Eq. 9.1}$$

Applying KCL at Node V_I:

$$\frac{V_I}{10} + \frac{V_I}{10} + \frac{V_I - V}{10} = 0 \Rightarrow V - 3V_I = 0 \qquad \text{Eq. 9.2}$$

Combining Eq. 9.1 and Eq 9.2 we obtain $R_{TH} = \frac{V}{I} = 30\Omega$.

(Thevenin's theorem)

10.

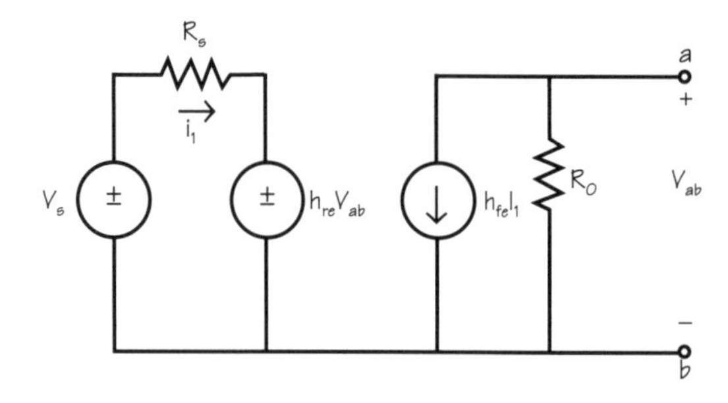

At the input: $V_s = I_1 R_s + h_{re} V_{ab} \Rightarrow I_1 = (V_s - h_{re} V_{ab})/R_s$
At the output (writing KCL): $h_{fe} I_1 + V_{ab}/R_0 = 0 \Rightarrow V_{ab} = V_{th} = -h_{fe} R_0 V_s/(R_s - R_0\, h_{fe} h_{re})$.

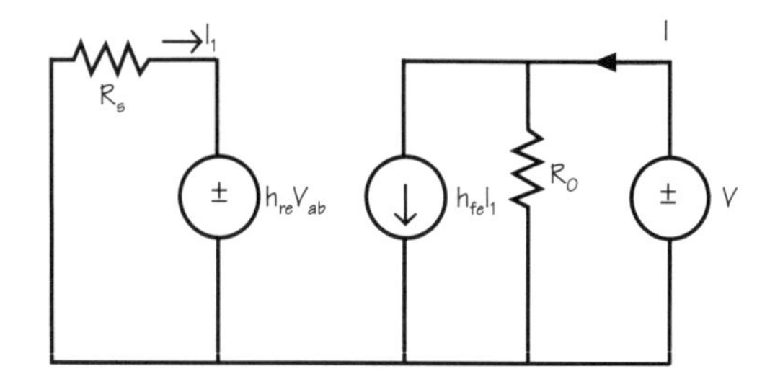

R_{th} is calculated based on the following circuit:

At the input: $hre\ V = -R_s I_1 \Rightarrow I_1 = -hre\ V/R_s$.
At the output side: $I = V/R_0 - hfe\ hre\ V/R_s \Rightarrow R_{th} = V/I = R_0 R_s/(R_s - R_0\ hfe\ hre)$

The resulting Thevenin's equivalent circuit is:

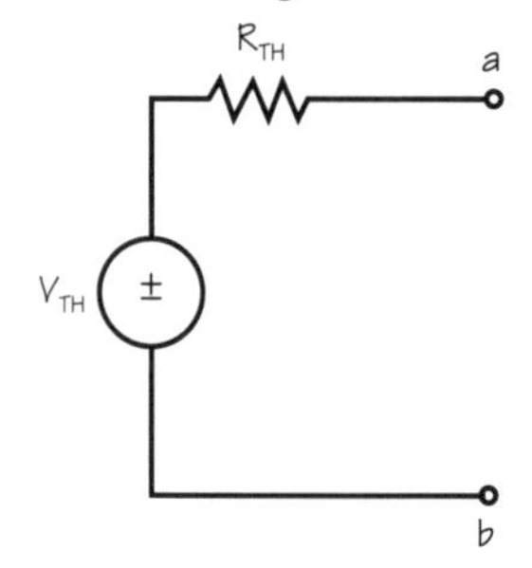

(Thevenin's theorem)

11.

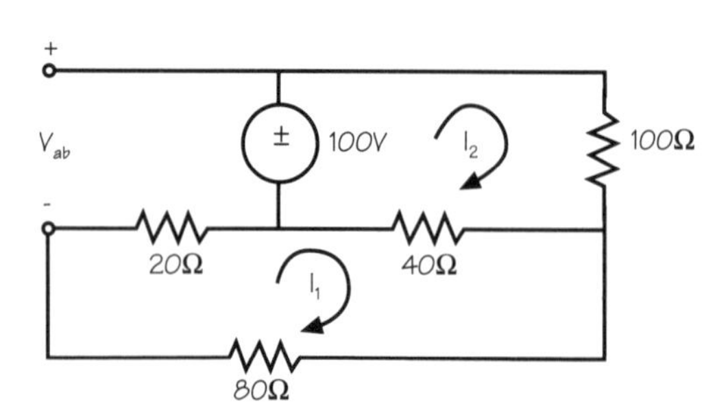

Method of mesh currents is used in this analysis.

$100 = 140\ I_2 - 40\ I_1$
$0 = -40\ I_2 + 140\ I_1$

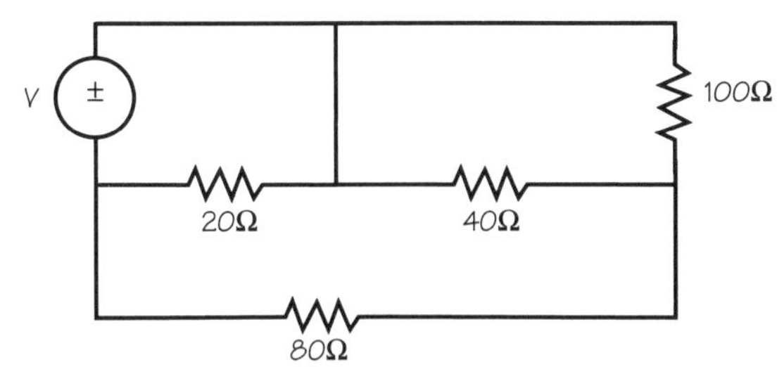

$\Rightarrow I_1 = 2/9$ A. Applying KVL: $V_{ab} = V_{th} = 100 - 20I_1 \Rightarrow V_{ab} = V_{th} = 95.56$

$$R_{TH} = \frac{V}{I} = 16.89$$

(Thevenin's theorem)

12. Method of mesh currents is used in this analysis.

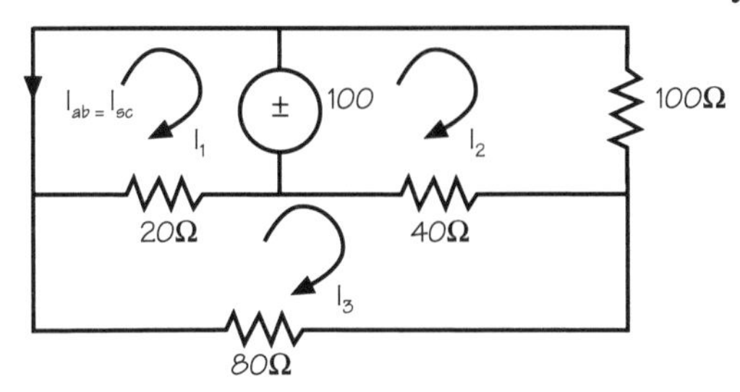

$$-100 = 20\,I_1 - 20\,I_3$$
$$100 = 140\,I_2 - 40\,I_3$$
$$0 = 140\,I_3 - 20\,I_1 - 40\,I_2$$

$$I_1 = \frac{\begin{pmatrix} -5 & 0 & -1 \\ 5 & 7 & -2 \\ 0 & 2 & 7 \end{pmatrix}}{\begin{pmatrix} 1 & 0 & -1 \\ 0 & 7 & -2 \\ -1 & -2 & 7 \end{pmatrix}} = -210/38 = : I_{sc} = I_{ab} = -I_1$$

R_{th} calculation is the same as shown in problem 12.11. **(Norton's theorem)**

13. For maximum power transfer to occur, the load resistance must be equal to the line resistance. From Thevenin's equivalent circuit the line resistance happens to be R_{th}. For maximum power transfer $R_L = R_{th} = 16.89\Omega$. **(Maximum power transfer)**

14. a) The 10V source is considered and the 50V source is replaced by its internal resistance to obtain the following circuit. By using series/parallel combination of resistors, the circuit is reduced as shown.

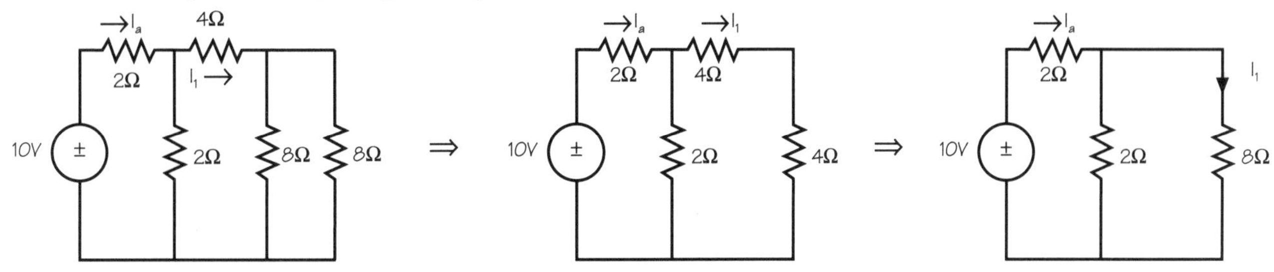

$I_A = 10/3.6 = 2.78$A. $I_1 = 2.78 * (2/10) = 0.556$A.

b) The 10V source is replaced by its internal resistance, and only the 50V source is considered.

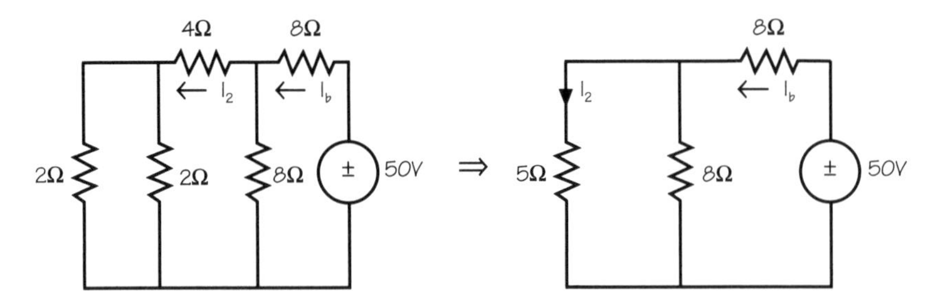

$I_B = 50/(8 + 8||5) = 4.5A$. $I_2 = 4.5 * (8/13) = 2.78A$.

The current I is the algebraic summation of I_1 and I_2. $I = I_1 - I_2 = 0.556 - 2.78 = -2.22A$.

(Superposition)

15. a)

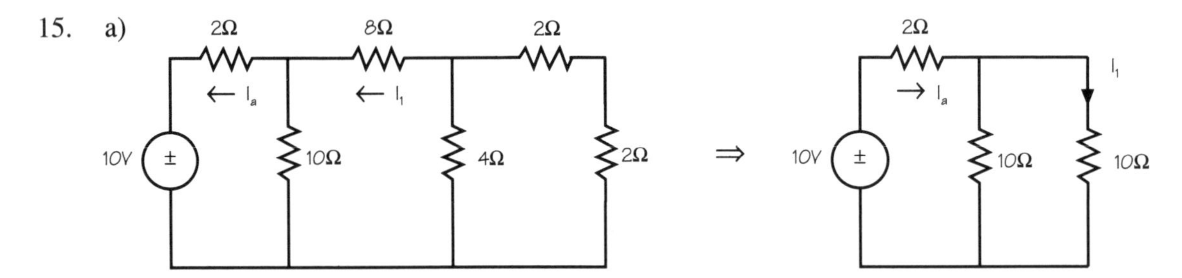

$I_A = 10/7 = 1.43A$. $I_1 = 1.43/2 = 0.71A$.

b)

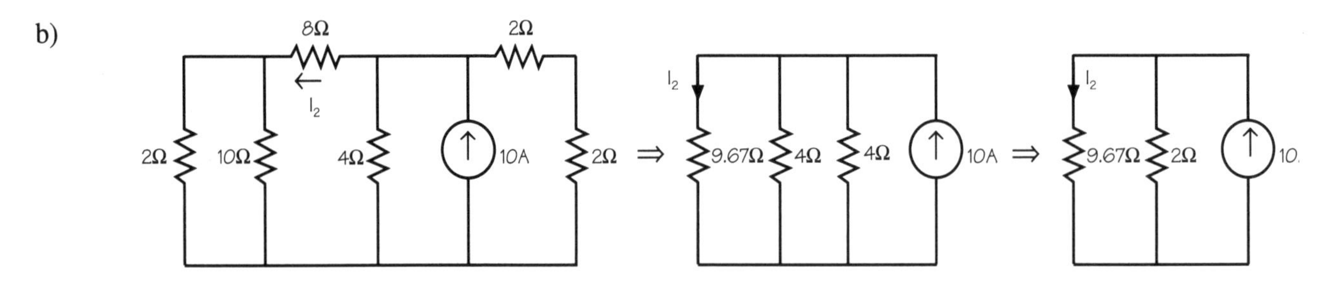

$I_2 = 10 * 2/(2 + 9.67) = 1.714A$.

c)

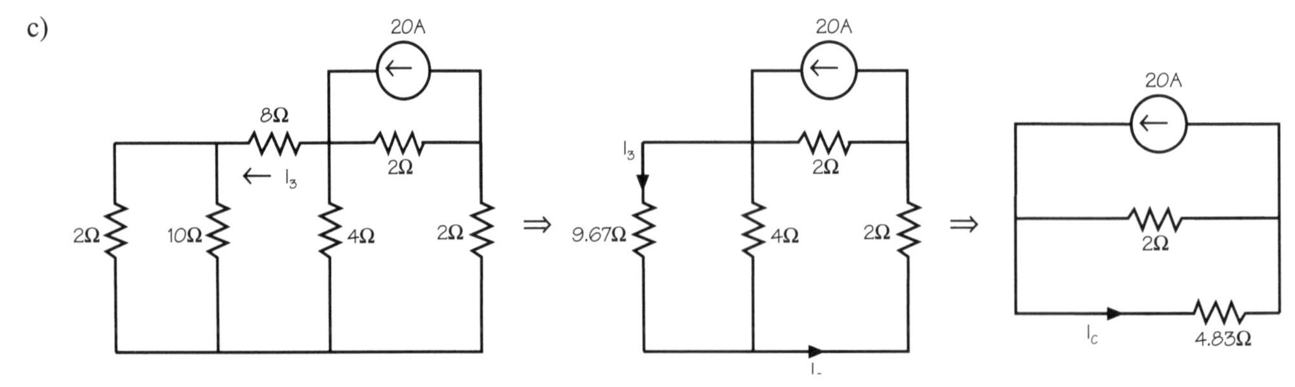

$I_C = 20 * 2/(2 + 4.83) = 5.86A$. $I_3 = 5.86 * 4/(4 + 9.67) = 1.714$ A.

The current I flowing through the 8Ω resistor is: $I = I_1 - I_2 - I_3 = 0.71 - 1.714 - 1.714 = -2.71A$.

(Superposition)

16. a)

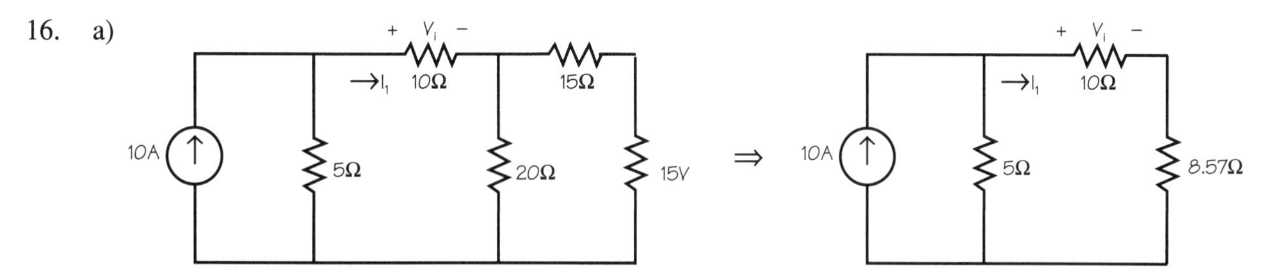

$I_1 = 10 * 5/23.57 = 2.12A$. $V_1 = 10 * 2.12 = 21.2V$.

b)

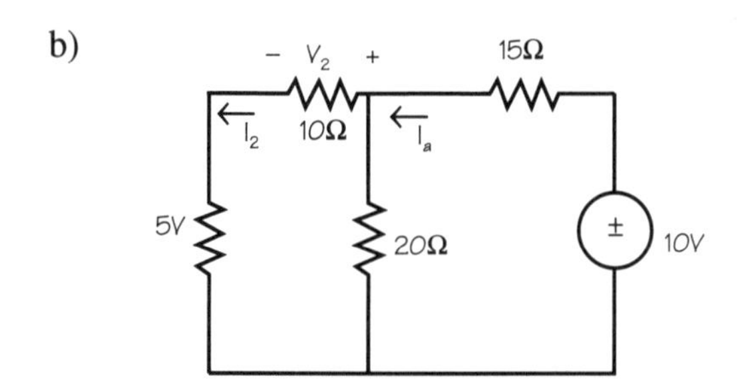

$I_a = 10/(15 + 8.57) = 0.424A$. $V_2 = 0.424 * 10 * 20/(20 + 15) = 2.424V$.

The voltage V is the algebraic summation of V_1 and V_2: $V = V_1 - V_2 = 18.78V$.
Note: The above analysis can also be carried out using the method of node voltages (as an example) yielding V_1 and V_2 directly. **(Superposition)**

17. (A voltage source with a series resistance is converted to a current source with a parallel resistance and vice versa).

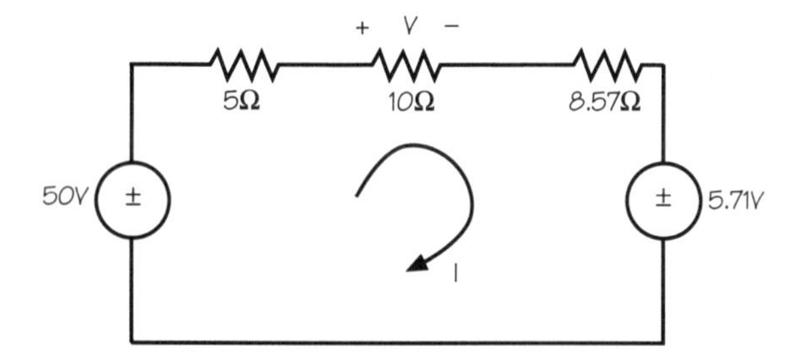

Transforming both the current and voltage source once we obtain:

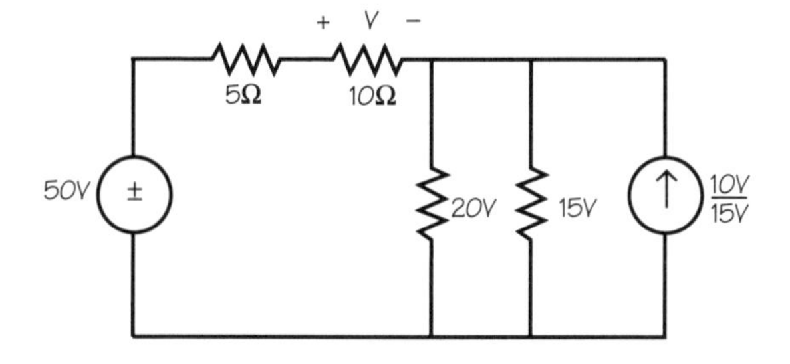

Transforming the current source into a voltage source, we obtain a simple series circuit.

$I = (50 - 5.71)/(5 + 10 + 8.57) = 1.88A$. $V = 1.88 * 10 = 18.8V$.

(Source transformation)

Grade Yourself

Circle the numbers of the questions you missed, then fill in the total incorrect for each topic. If you answered more than three questions incorrectly, you need to focus on that topic. (If a topic has less than three questions and you had at least one wrong, we suggest you study that topic also. Read your textbook, a review book, or ask your teacher for help.)

Subject: Network Theorems

Topic	Question Numbers	Number Incorrect
Mesh current	1, 2, 3, 4	
Nodal analysis	5, 6, 7	
Thevenin's theorem	8, 9, 10, 11	
Norton's theorem	12	
Maximum power transfer	13	
Superposition	14, 15, 16	
Source transformation	17	

Circuit Transients

Brief Yourself

Switching transients in electric circuits is a very important phenomenon, since the total response is always a summation of the steady-state and transient components. In this chapter, transient in simple RLC circuits under dc excitation is investigated. Storage elements, such as inductors and capacitors, play a major role in determining the initial conditions used in the calculation of transient response. The voltage drop across a capacitor cannot change instantaneously, unless the change in current at that instant of time is infinitely large. For an inductor, the current flowing through it cannot change instantaneously unless the change in voltage across it at that instant is infinite.

Test Yourself

1. For the RL circuit shown, the switch is shifted from position *a* to position *b* at t = 0. Calculate i(t) for $t \geq 0$.

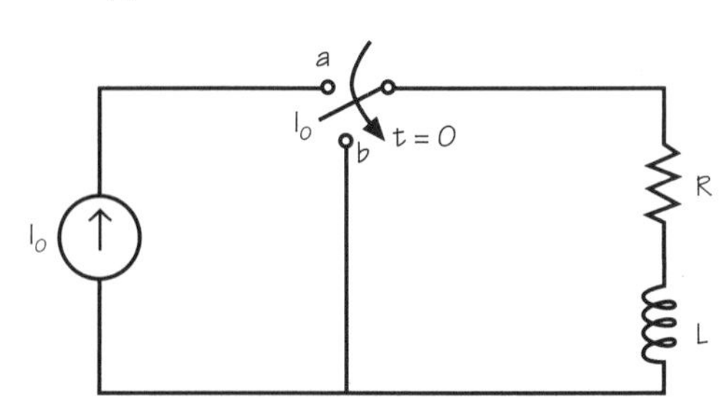

2. Using the above circuit, calculate: a) the voltage across the circuit elements; b) the power dissipated by the resistor; and c) the energy stored by R and L.

3. For the circuit shown calculate: a) the current i(t) flowing through the inductance; b) voltage drop across the resistor; c) voltage drop across the inductor; and d) power dissipated by the resistor.

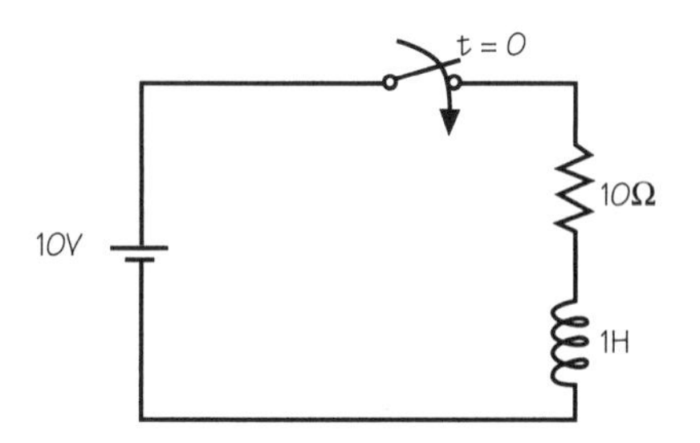

4. For the RL circuit shown, calculate $v_L(t)$ for $t > 0$.

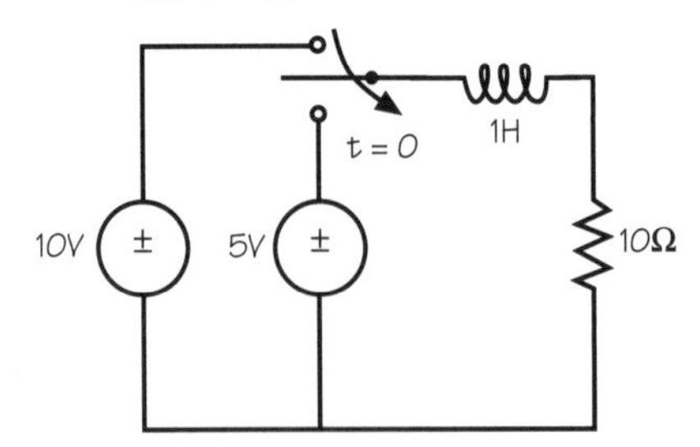

5. In the RC circuit shown, the switch was at position 1, and then at $t = 0$ it was switched to position 2. Calculate $v_C(t)$.

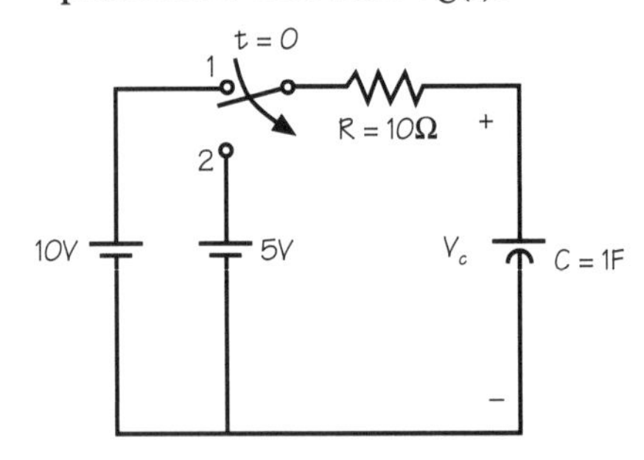

6. For the circuit shown, calculate $v_C(t)$.

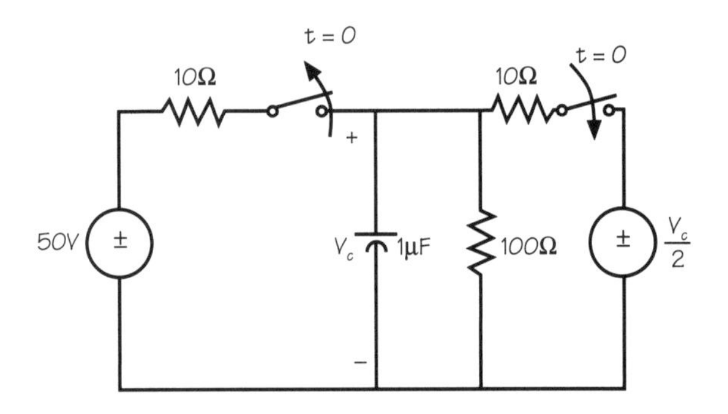

7. For the circuit, determine $v_C(t)$ for $t > 0$.

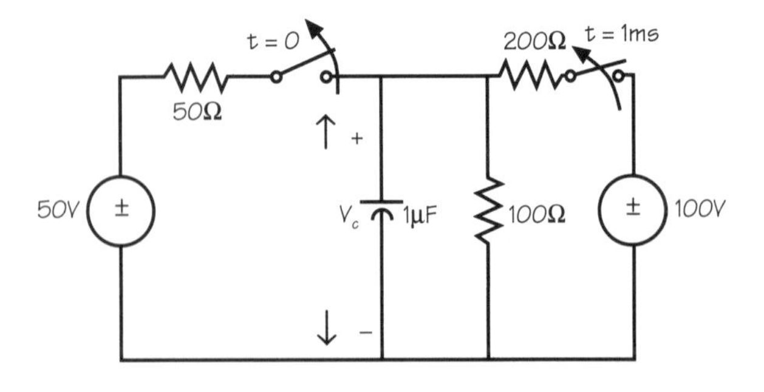

8. The switch is flipped from position *A* to position *B* at $t = 0$. Calculate $v_C(t)$ for $t > 0$.

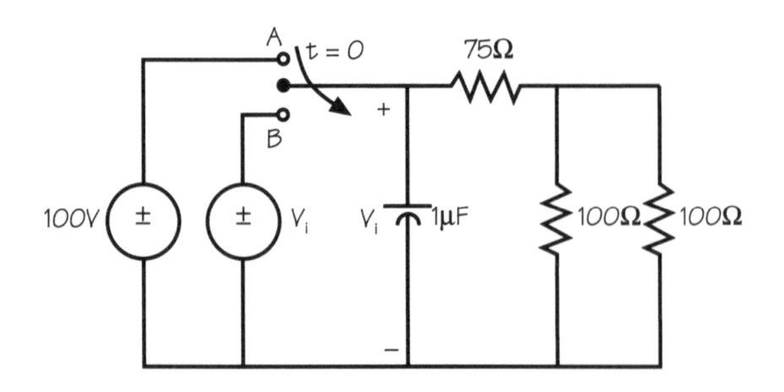

9. For the following RLC circuit, calculate the current through the inductance for $t > 0$.

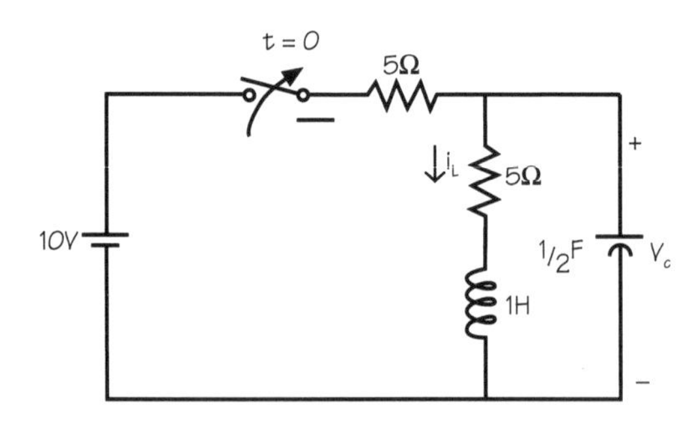

10. Calculate the voltage drop across the capacitor, $v_C(t)$, for $t > 0$.

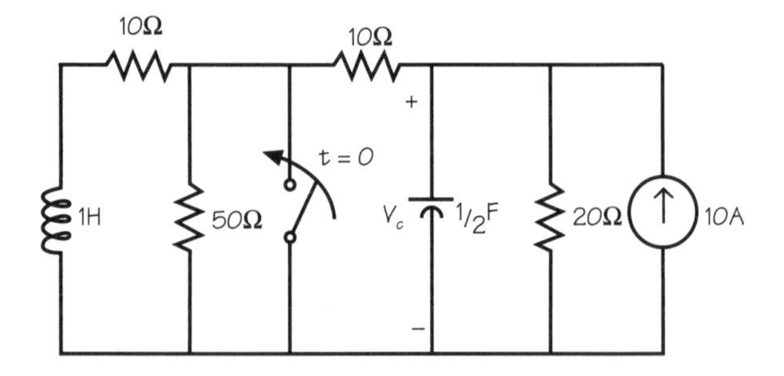

Check Yourself

1.

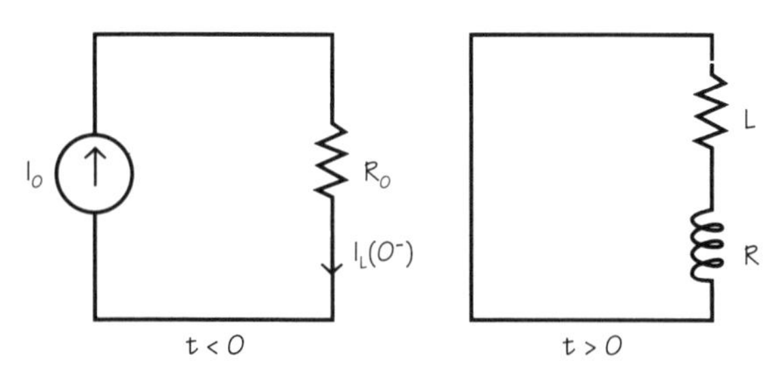

For $t < 0$, the current through the inductance $i_L(0-) = I_0$.

For $t \geq 0$, the governing equation is obtained by writing the KVL:

$$L\frac{di}{dt} + Ri = 0; \frac{di}{dt} + (R/L)\ i = 0;$$

The solution of this first-order differential equation can be written as (by using the method of integrating factors or other simple methods):

$$i(t) = C \exp(-Rt/L).$$

As the current through the inductance cannot change instantaneously $i_L(0+) = i_L(0-) = I_0$. Implying $I_0 = C$. Therefore, $i(t) = I_0 \exp(-Rt/L)$. [*Note:* This particular equation shows how current in a sourceless R-L circuit decays. The circuit had some initial condition (the inductance had some current I_0 when the switch was connected to terminal b; energy is stored in the form of magnetic field in inductances). The decay is exponential in nature, with a characteristic time constant $\tau_{LR} = L/R$ sec.] **(RL transients)**

2. a) The voltage across the resistor, $v_R(t) = i(t)\ R$; $v_R(t) = R\ I_o \exp(-Rt/L)$.

   Voltage across the inductance $v_L(t) = L\frac{di}{dt}$; $v_L(t) = -R\ I_o \exp(-Rt/L)$. [Note: $v_L(t) + v_R(t) = 0$, as is suggested by KVL]

   b) Power dissipated by the resistor: $P_R(t) = i^2(t)\ R = v_R(t)\ i(t) = R\ I_o^2 \exp(-2Rt/L)$.

   c) Total energy dissipated by R: $W_R = RI_o^2 \int_0^{\infty} \exp(-2Rt/L)dt = (1/2)\ LI_o^2$.

   Energy stored in the inductance $W_L = (1/2)\ LI_o^2$.

   [*Note:* The total energy that was stored in the form of magnetic field in the inductance is eventually dissipated by the resistor. At $t = \infty+$, the circuit has lost all of its energy and the current flowing through the circuit is identically zero. **(RL transients)**

3. a) $t < 0$: the circuit was initially de-energized, implying $i(0-) = 0$.

   b) $t > 0$.

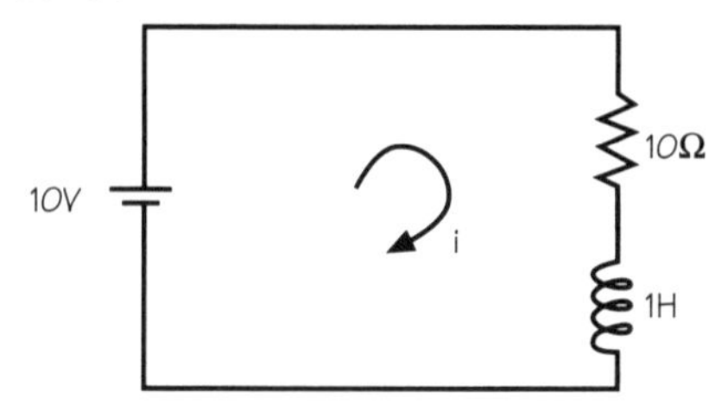

The KVL is written as:

$$\frac{di}{dt} + 10i = 10.$$

Integrating factor = exp(10t), which, after integration, will yield: $i(t) = 1 + C\exp(-10t)$, where C is an arbitrary constant. C is evaluated by considering the initial condition: $i(0+) = i(0-) = 0$ (current flowing through an inductance can not change instantaneously). Therefore, $i(t) = 1 - \exp(-10t)$.

$v_R(t) = i(t) * R = 10(1-\exp(-10t))$

c) $v_L(t) = \frac{di}{dt} = 10\exp(-10t)$

d) $p_R(t) = i^2 R = 10(1 - 2\exp(-10t) + \exp(-20t))$. **(RL transients)**

4. $t < 0$:

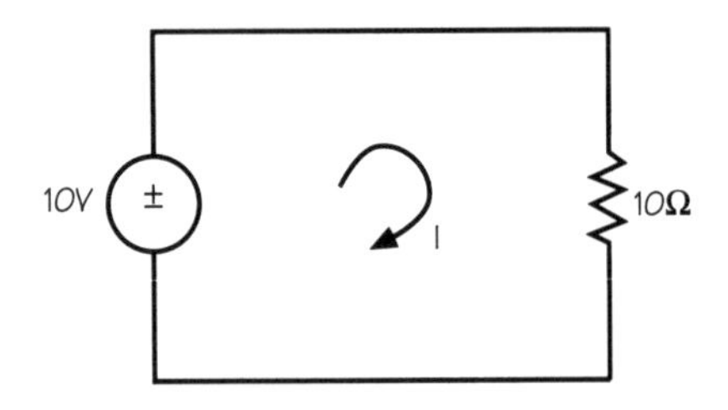

$I = 1A$; $i(0-)=1A$.

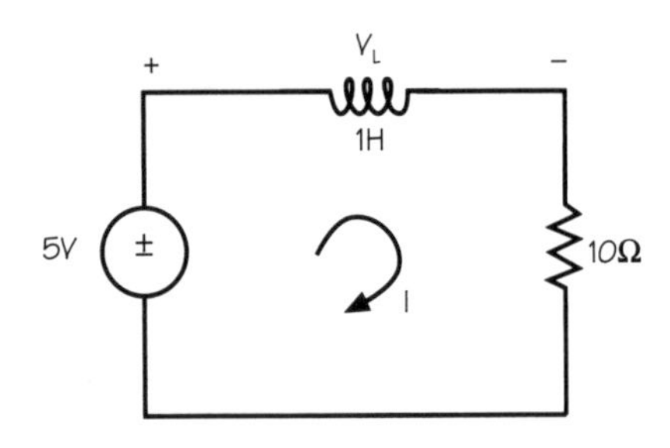

$t > 0$:

$5 = \frac{di}{dt} + 10i(t)$. [Using an integrating factor of exp(10t)], the differential equation can be written as:

$$\int d(ie^{10t}) = 5\int e^{10t}dt + C \Rightarrow i(t) = 0.5 + C\exp(-10t) \text{ (where C is the arbitrary constant).}$$

Now, as the current through an inductance cannot change instantaneously: $i(0+) = i(0-) = 1A$. Implying $C = 0.5$

$i(t) = 0.5(1 + \exp(-10t))$. Therefore, $v_L(t) = \frac{di}{dt} = -5\exp(-10t)$.

(RL transients)

5. $t < 0$:

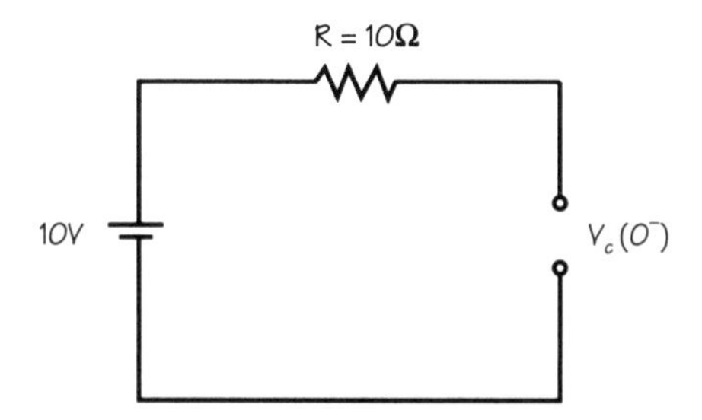

The capacitor will be charged to 10V and will behave like an open circuit.

$v_C(t = 0-) = 10V.$

$t > 0$:

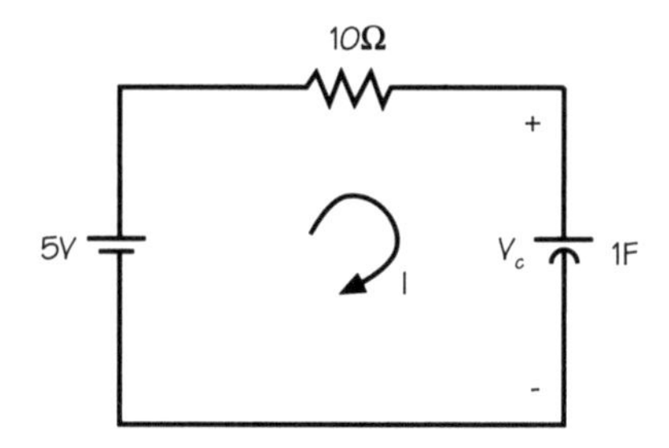

Applying KCL: $(5 - v_C)/10 = C\frac{dv_C}{dt} \cdot \frac{dv_C}{dt} + 0.1v_C = 0.5.$
Integrating factor = exp(0.1t). Using the integrating factor, the above differential equation will yield: $v_C(t) = 5 + A\exp(-0.1t)$ (where A is the arbitrary constant). Now, the voltage across a capacitor cannot change instantaneously $\Rightarrow v_C(0-) = v_C(0+) = 10V$. Giving us,

$A=5$. $v_C(t) = 5(1 + \exp(-0.1t))$ **(RC transients)**

6. $t < 0$:

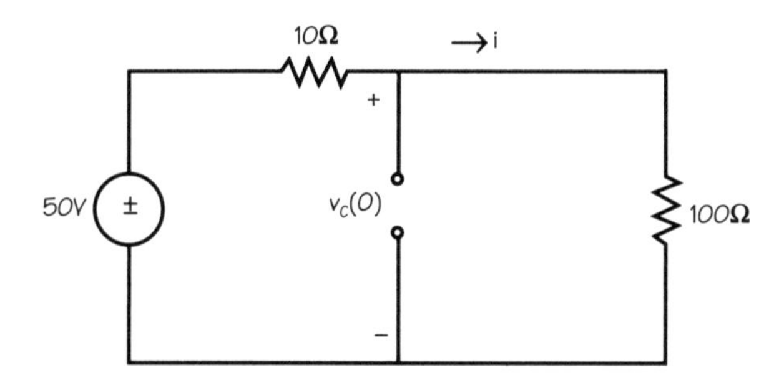

$i = 50/(110)$ A. $v_C(0-) = (50/110) * 100 = 500/11$ V.

$t > 0$:

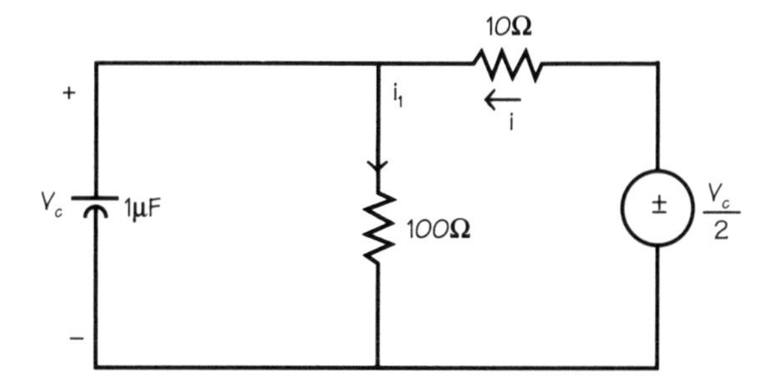

Writing the node voltage equation at node A:

$(v_C/2 - v_C)/10 = v_C/100 + C\frac{dv_C}{dt}$. That reduces to $\frac{dv_C}{dt} = -6*10^4\ v_C$. Implying

$v_C(t) = B\ \exp(-6 * 10^4 t)$.

Now $v_C(0-) = v_C(0+) = 500/11$. Therefore, $v_C(t) = (500/11)\ \exp(-6 * 10^4 t)$. **(RC transients)**

7. Initial condition (t < 0)

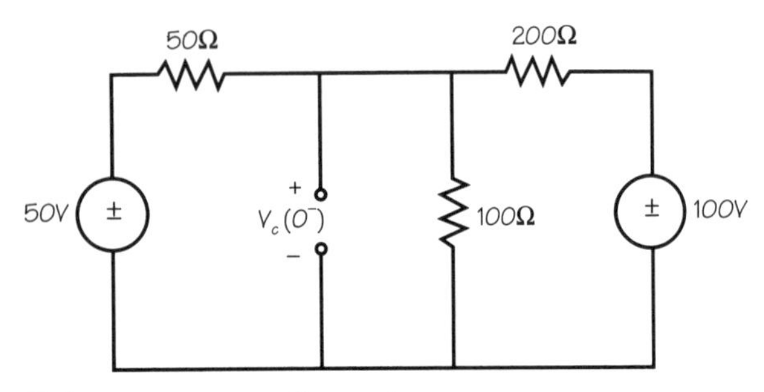

Using source transformation:

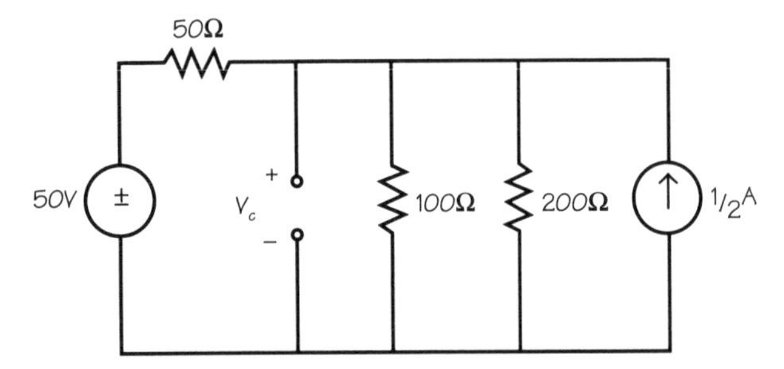

Source transforming the 1/2A source back to a voltage source:

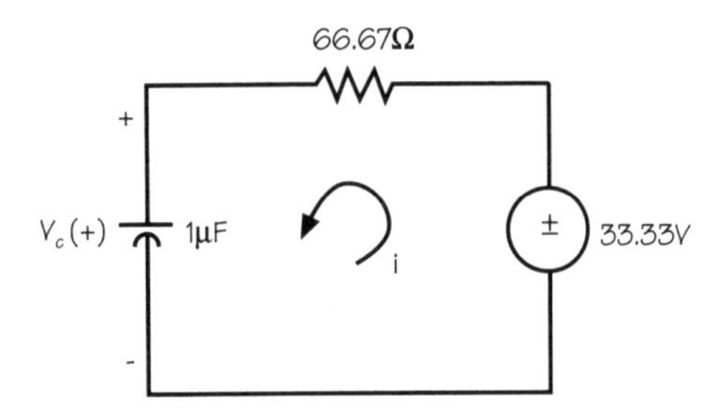

$i = (50-33.33)/(50 + 66.67) = 0.143.$

$v_C(0-) = 50 - 50 * .143 = 42.86$ V

0 < t < 1ms:

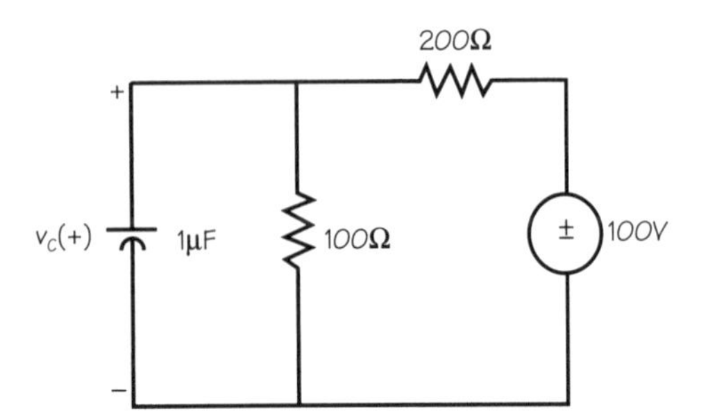

Using source transformation, the above circuit can be redrawn as follows, yielding:

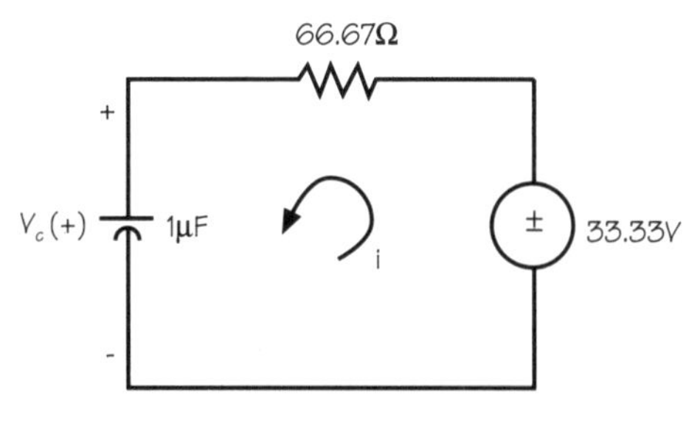

$$66.67C\frac{dv_C}{dt} + v_C(t) = 33.33$$

$$\frac{dv_C}{dt} + 15000\ v_C = 5*10^5$$. After integration (using integrating factor)

$$v_C(t) = 33.33 + A\exp(-1.5*10^4\ t).$$

As the voltage across any capacitor cannot change instantaneously: vc(0–) = vc(0+) = 42.86V, implying A = 9.5.

$$v_C(t) = 33.33 + 9.5\exp(-1.5 * 10^4\ t).$$
$$v_C(t = 1\text{ms}) = 33.33\text{ V}.$$

t > 1ms:

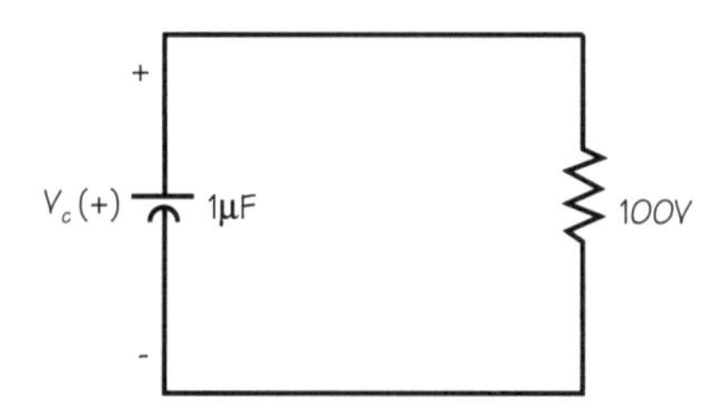

$\tau_{RC} = RC = 0.1$ms.
$v_C(t) = D\exp(-(t-1\text{ms})/(\tau_{RC})$
$v_C(1\text{ms}-) = v_C(1\text{ms}+) = 33.33$V. Therefore, D = 33.33V.
$v_C(t) = 33.33\exp(-10^4\ (t-1\text{ms}))$.

Note: This particular RC circuit has two time constants and depends upon the circuit configuration.

(RC transients)

8. Initial condition (t < 0):
   $v_C(0-) = 100 - 100(100/225) = 55.56$V.

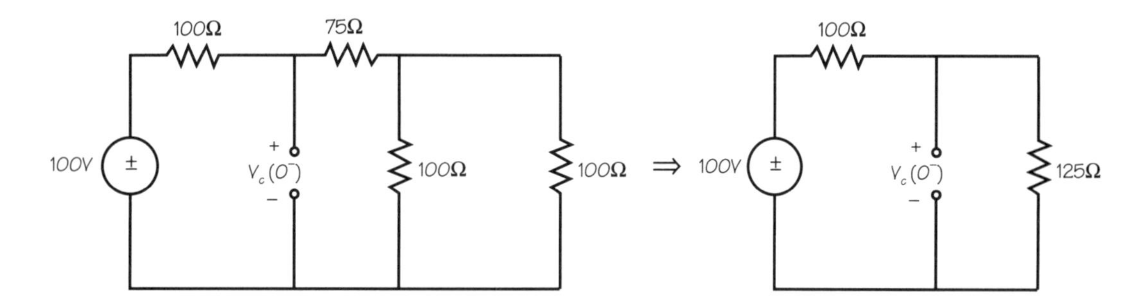

t > 0:

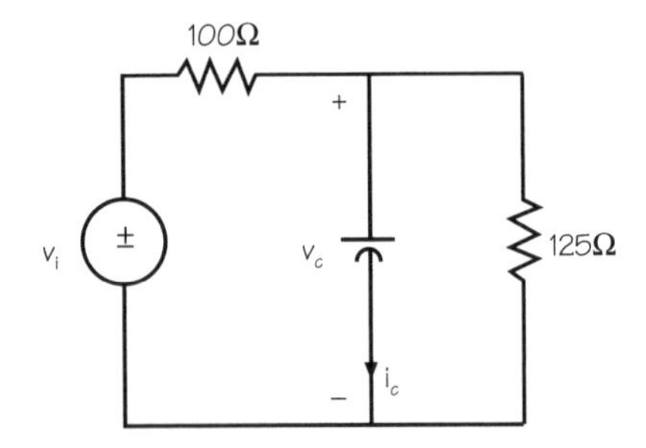

Using KCL: $(v_i - v_C)/100 = i_C + v_C/125$; where ic the capacitor current and equals $C\frac{dv_C}{dt}$.

The above equation can be recast as: $\frac{dv_C}{dt} + 1.8*10^4\ v_C = 10^4\ v_i$.

For $0 < t < 1\mu s$: $v_C(t) = 0.56 + D\exp(-1.8 * 10^4 t)$. Where D = 55V.

$v_C(t) = 0.56 + 55\exp(-1.8 * 10^4 t)$.
$v_C(t = 1\mu s) = 54.57$ V.

For $t > 1\mu s$: $v_C(t) = E\exp(-1.8 * 10^4(t - 1\mu s))$, where E = 54.57.
$v_C(t) = 54.57\exp(-1.8 * 10^4\ (t - 1\mu s))$.

(RC transients)

9. Initial condition:

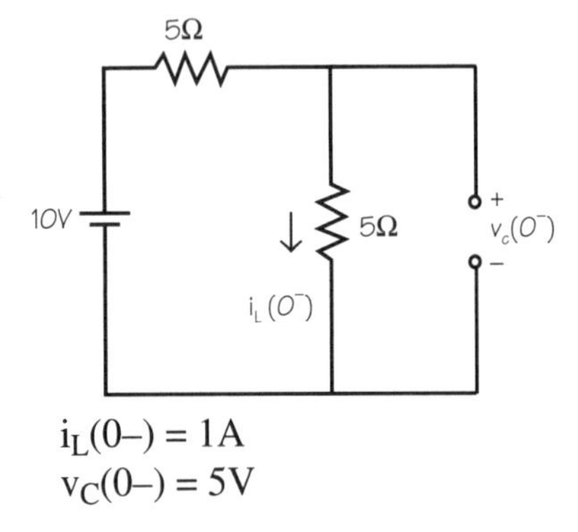

$i_L(0-) = 1A$
$v_C(0-) = 5V$

t > 0:

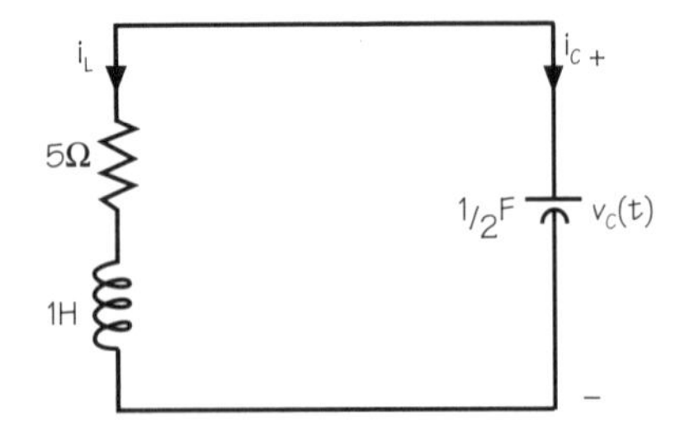

Applying KCL: $i_L(t) + i_C(t) = 0$; where $i_C(t) = \frac{Cdv_C}{dt} \cdot v_C(t) = 5\ i_L + L\frac{di}{dt}$.

The governing equation, therefore, becomes:

$$\frac{d^2i_L}{dt^2} + \frac{5}{L}\frac{di_L}{dt} + \frac{1}{LC}i_L(t) = 0; \text{ that can be rewritten as } \frac{d^2i_L}{dt^2} + \frac{di_L}{dt} + 2i_L(t) = 0.$$

The characteristic equation is: $s^2 + 5s + 2 = 0$, and the roots are found to be

$s_{1,2} = -6.623, 1.623.$
$i_L(t) = A \exp(-6.623t) + B \exp(1.623t).$

Evaluation of the arbitrary constants:

$i_L(0-) = i_L(0+) = 1. \Rightarrow A + B = 1.$

Also, $v_C(0-) = v_C(0+) = 5\, i_L(0+) + L\frac{di}{dt}(t = 0+).$

$5i(0+) = 5V \cdot v_L(0+) = -6.623A + 1.623B. \Rightarrow A = 0.2$ and $B = 0.8.$
$i_L(t) = 0.2 \exp(-6.623t) + 0.8 \exp(1.623t)$

Note: The system is overdamped. **(RLC transients)**

10. Initial condition ($t < 0$):
    $i = 200/(30 + 25/3) = 5.2\text{A}.$
    $i_L(0-) = 5.2 - 50/60 = 4.35$ A.
    $v_C(0-) = 200 - 20 * 5.2 = 96\text{V}.$

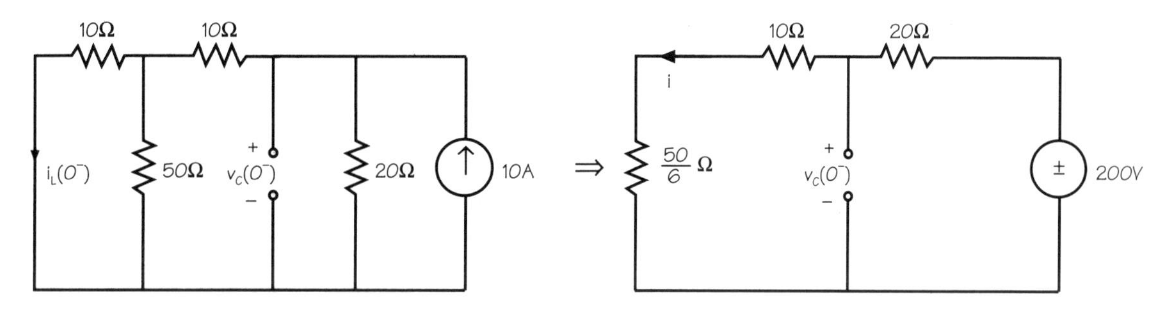

$t > 0$:

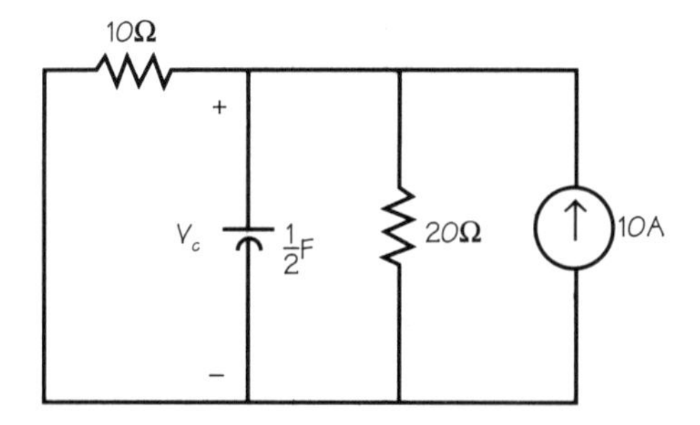

For the RC circuit with a source, $v_C(t)$ can be written as a combination of forced and natural response.

a) Forced Response:

$I_1 = 10 * (20/30) = 6.67\text{A}.$
$v^F_C = 10 * 6.67 = 66.7$ V

b) Natural Response:

$\tau_{RC} = RC = 3.33$
$v^N_C(t) = A \exp(-0.3t)$

c) Total Response:

The total response: $v_C(t) = v^F_C + v^N_C(t)$

$v_C(t) = 66.7 + A \exp(-0.3t)$.

Using $v_C(0-) = v_C(0+) = 96 = 66.7 + A \Rightarrow A = 29.33$.

$v_C(t) = 66.7 + 29.33 \exp(-0.3t)$ V.

(RLC transients)

Grade Yourself

Circle the numbers of the questions you missed, then fill in the total incorrect for each topic. If you answered more than three questions incorrectly, you need to focus on that topic. (If a topic has less than three questions and you had at least one wrong, we suggest you study that topic also. Read your textbook, a review book, or ask your teacher for help.)

Subject: Circuit Transients

Topic	Question Numbers	Number Incorrect
RL transients	1, 2, 3, 4	
RC transients	5, 6, 7, 8	
RLC transients	9, 10	

AC Circuits

Brief Yourself

The use of network theorems to solve dc circuits is extended to ac circuits. Network theorems are used to solve for the steady-state response of ac circuits. The total response is always the sum of the steady-state and the transient response.

Test Yourself

1. For the RLC circuit shown, calculate the total response for $t \geq 0$.

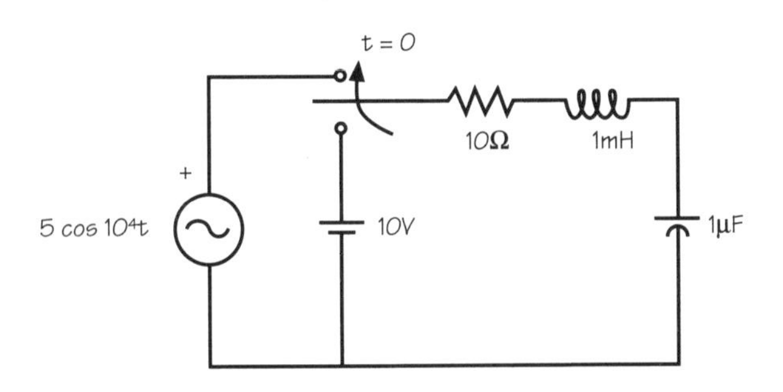

2. For the circuit shown, calculate the output voltage.

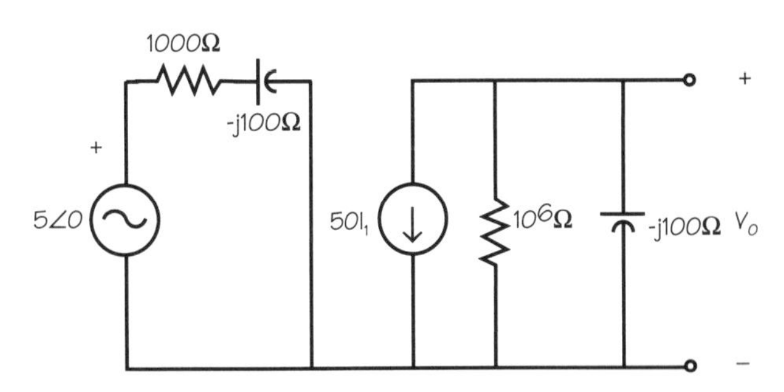

3. Using mesh currents, solve for Vc.

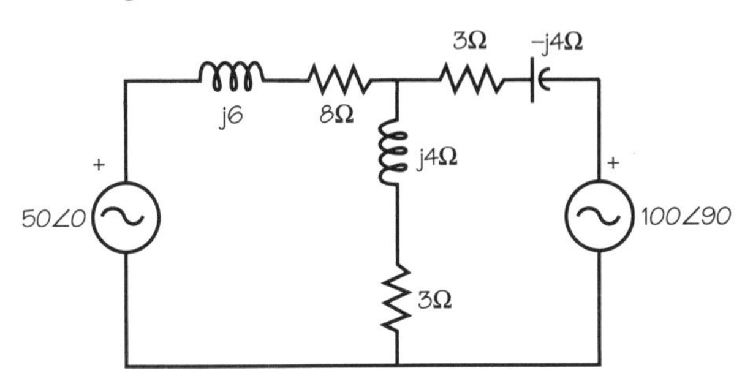

4. For the circuit shown, calculate the current through the inductance.

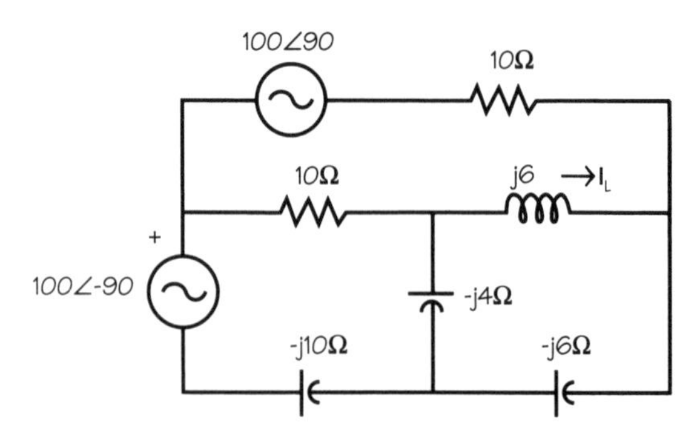

5. Using the method of node voltages, calculate the current through the 50∠0 V source.

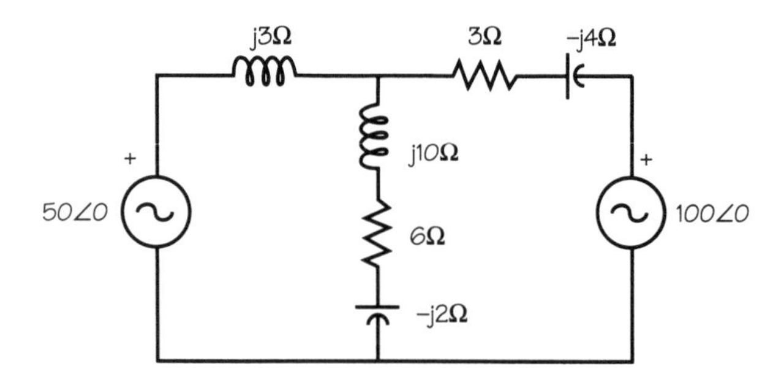

6. For the circuit shown, calculate all the node voltages.

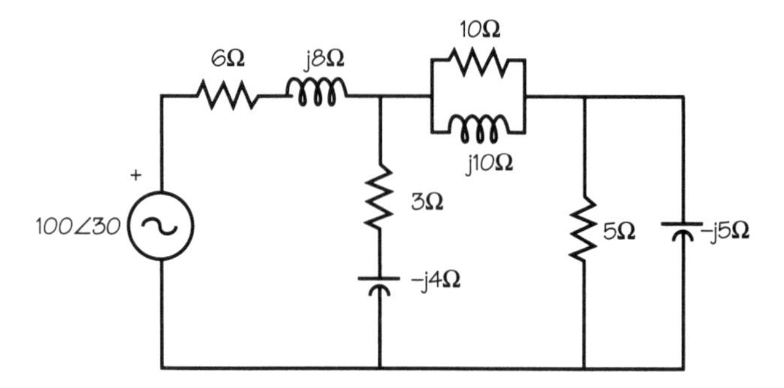

7. For the circuit shown, determine the current through the independent source.

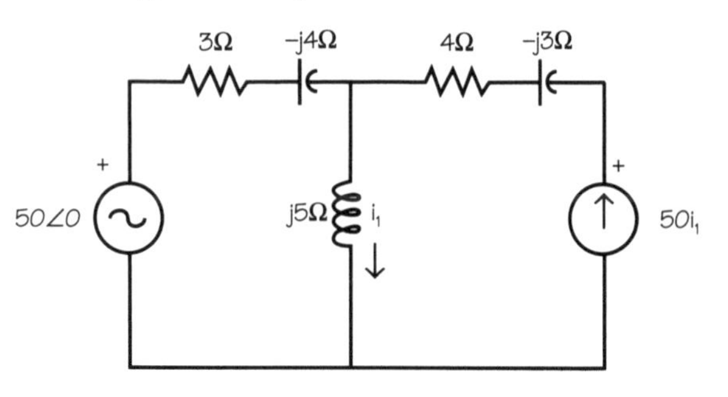

8. Calculate the voltage difference between the designated nodes.

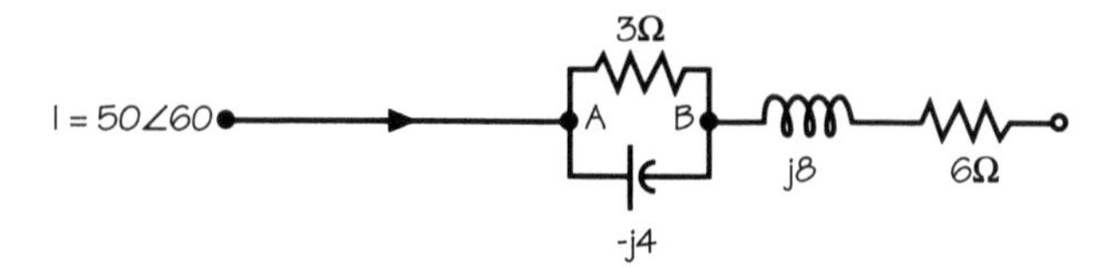

9. Replace the bridge circuit by looking into the terminals *a-b* by its Thevenin's equivalent.

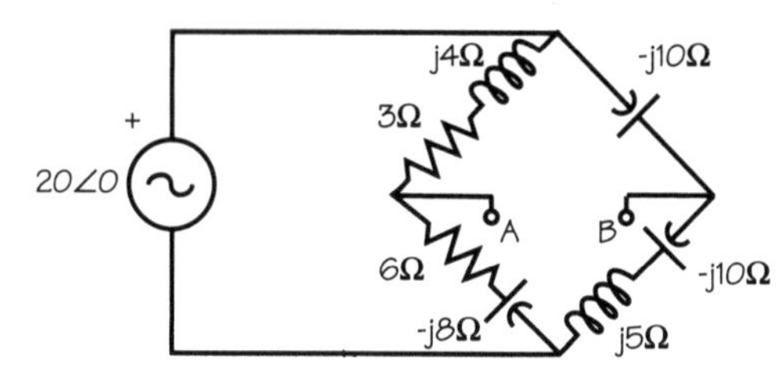

10. For the circuit shown, calculate the current I, by first replacing the circuit by its Thevenin's equivalent.

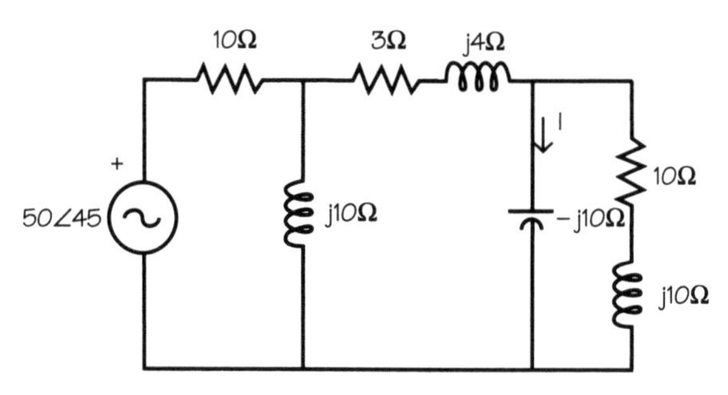

11. Determine the load impedance to be connected at terminals *a-b*, in the bridge circuit shown in problem 4.9, for maximum power transfer.

12. For the circuit shown, calculate the current I, using the method of superposition.

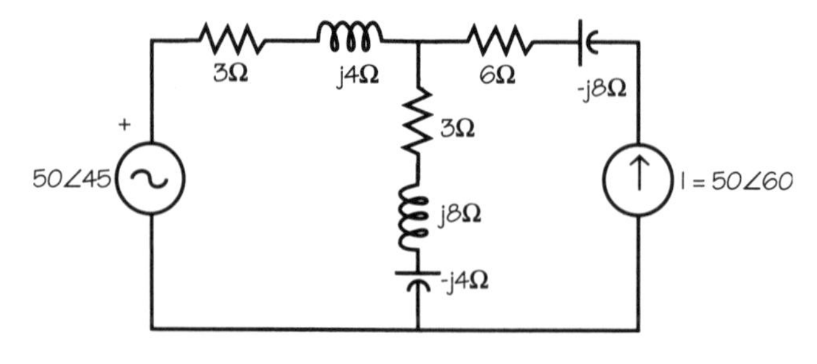

Check Yourself

1. $t < 0$

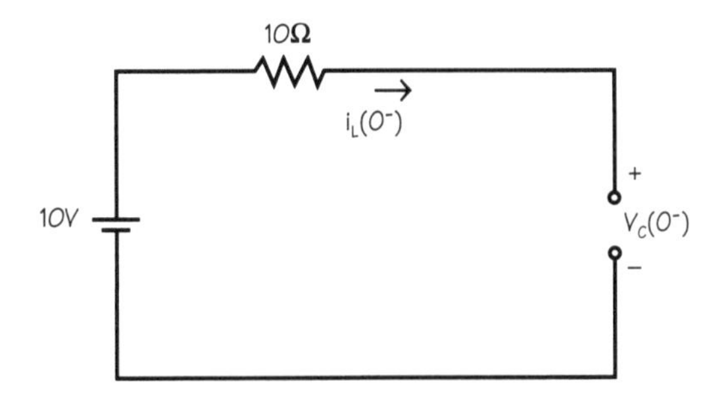

$v_C(0-) = 10V$
$i_L(0-) = 0A$

$t > 0$: Steady State

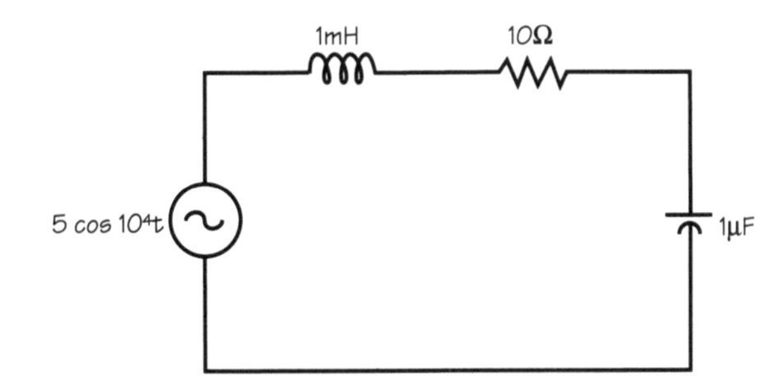

$\omega = 10^4$ rad/sec
$X_L = \omega L = 10\Omega$
$X_C = 1/\omega C = 100\Omega$

The above circuit can be redrawn as:

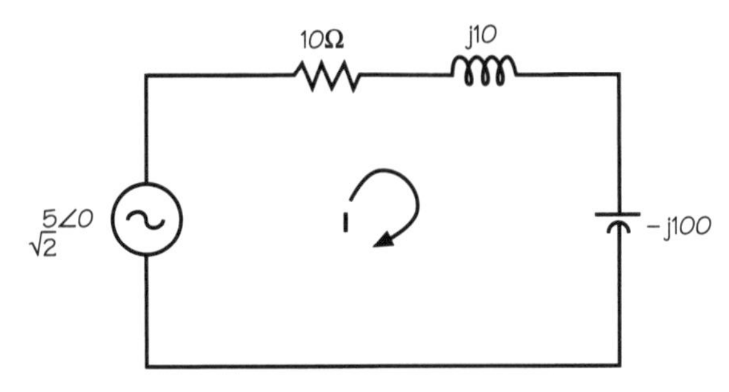

Using the method mesh currents: $\frac{5}{\sqrt{2}}\angle 0 = (10 + j10 - j\,100)\,\mathbf{I}$

$\mathbf{I} = \frac{5}{\sqrt{2}}\angle 0/(10 - j90) = \frac{55.2}{\sqrt{2}}\angle 83.66$ mA.

$= 39.032\angle 83.66$mA

Transient Response:

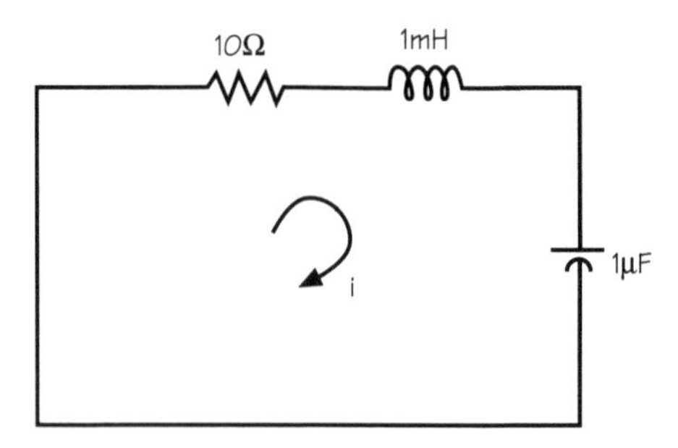

The governing equation is:

$$L\frac{di}{dt} + Ri + \frac{1}{C}\int idt = 0.$$ That reduces to

$$\frac{d^2i}{dt^2} + \frac{R}{L}\frac{di}{dt} + \frac{1}{LC}i = 0.$$

Inserting the magnitudes of R, L, and C, the above equation reduces to:

$$\frac{d^2i}{dt^2} + 10^4\frac{di}{dt} + 10^9 i = 0.$$

The characteristic equation is therefore: $s^2 + 10^4 s + 10^9 = 0$. The roots are: $s_{1,2} = -5 \times 10^3 \pm j3.12 \times 10^4$.

The transient response, therefore, can be written as:

$$i(t) = (A \cos 3.12 \times 10^4 t + B \sin 3.12 \times 10^4 t)\, e^{-5\times10^3 t}.$$

The total current:

$$i_T(t) = 55.2 \times 10^{-3} \cos(10^4 t + 83.66) + (A \cos 3.12 \times 10^4 t + B \sin 3.12 \times 10^4 t)\, e^{-5\times10^3 t}$$

Now, to evaluate the arbitrary constants, we make use of the initial conditions:

$$i(0^-) = i(0^+) = 55.2 \times 10^{-3} \cos(83.66) + A.$$ Implying, $A = -6.09 \times 10^{-3}$.

From KVL: $v_{in}(0^+) = v_R(0^+) + v_L(0^+) + v_C(0^+)$.
With, $v_R(0+) = Ri(0+) = 0$

$v_C(0^+) = v_C(0^-) = 10V$.
$v_{in}(0^+) = 5V$, implying $v_L(0^+) = 5 - 10 = -5V$.

Now, $v_L(0^+) = L\frac{di}{dt}(\text{at } t = 0^+) = 1 \times 10^{-3}\, [-548.6 - 5 \times 10^3 A + 3.12 \times 10^4 B]$.
Therefore, $B = -143.65 \times 10^{-3}$.
The total current can now be written as:

$$i_T(t) = 55.2 \times 10^{-3} \cos(10^4 t + 83.66) - 143.78\, e^{-5\times10^3 t} \sin(3.12 \times 10^4 t + 2.43) \text{ mA}.$$

(RLC circuit complete solution [steady-state plus transient])

2.

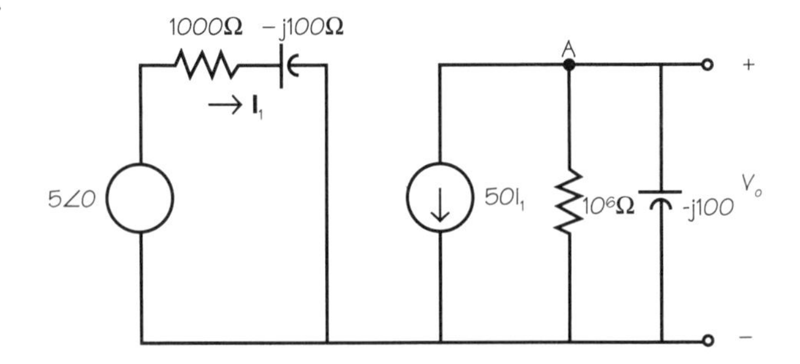

Using the method of mesh currents at the input:

$$5\angle 0 = (1000 - j100)\,\mathbf{I}_1\mathbf{:}\ [\mathbf{I}_1 = 4.97 \times 10^{-3}\angle 5.7\text{A.}]$$

Applying KCL at node A:

$$50\mathbf{I}_1 + \frac{\mathbf{V}_0}{10^6} + \frac{\mathbf{V}_0}{-j100} = 0.$$

Using $\mathbf{I}_1$, the output voltage can be expressed as:

$$\Rightarrow \mathbf{V}_0(10^{-6} + j10^{-2}) = -50\ * 4.97 \times 10^{-3}\angle 5.7$$

$\mathbf{V}_0 = -24.8\angle -84.3$ V. **(Mesh current)**

3. The mesh current equations are written as:

$$50\angle 0 = (3 + j4 + 8 + j6)\,\mathbf{I}_1 - (3+j4)\,\mathbf{I}_2$$
$$-100\angle 90 = -(3+j4)\,\mathbf{I}_1 + (3+j4 + 3-j4)\,\mathbf{I}_2.$$

Now, using Cramer's rule:

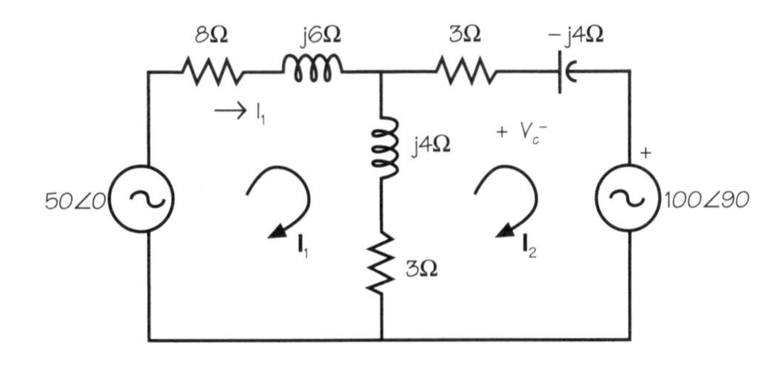

$$\mathbf{I}_2 = \frac{\begin{pmatrix} 11+j10 & 50 \\ -3-j4 & -j100 \end{pmatrix}}{\begin{pmatrix} 11+j10 & -3-j4 \\ -3-j4 & 6 \end{pmatrix}} = 17.94\angle -64.25.$$

Therefore, $\mathbf{V}_C = -j4 \times \mathbf{I}_2 = 71.75\angle -154.25$ V. **(Mesh current)**

4.

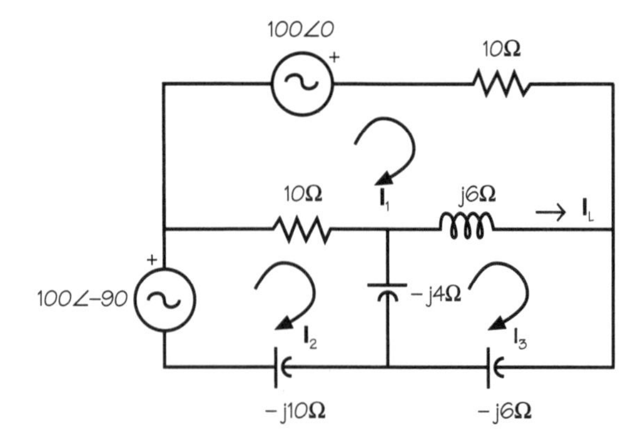

The mesh current equations can be written as:

$$100\angle 0 = (20 + j6)\,\mathbf{I}_1 - 10\,\mathbf{I}_2 - j6\,\mathbf{I}_3$$
$$100\angle{-90} = -10\,\mathbf{I}_1 + (10 - j14)\,\mathbf{I}_2 + j4\,\mathbf{I}_3$$
$$0 = -j6\,\mathbf{I}_1 + j4\,\mathbf{I}_2 - j4\,\mathbf{I}_3.$$

Using the last equation, the first two equations can be simplified to:

$$\begin{pmatrix} 20 + j15 & -j16 \\ -10 - j6 & 10 - j10 \end{pmatrix}\begin{pmatrix} \mathbf{I}_1 \\ \mathbf{I}_2 \end{pmatrix} = \begin{pmatrix} 100 \\ -j10 \end{pmatrix}.$$

Using Cramer's rule:

$$\mathbf{I}_1 = \frac{\begin{vmatrix} 100 & -j16 \\ -j100 & 10 - j10 \end{vmatrix}}{\begin{vmatrix} 20 + j15 & -j6 \\ -10 - j6 & 10 - j10 \end{vmatrix}} = 2.366\angle 56.17.$$

And $\mathbf{I}_2 = 5.81\angle{-4}$ A.

Now $\mathbf{I}_3 = \mathbf{I}_2 - 1.5\,\mathbf{I}_1 = 5.086\angle{-41.32}$ A.
The current through the inductance $I_L = \mathbf{I}_3 - \mathbf{I}_1 = 5.8817\angle{-64.85}$ A. **(Mesh current)**

5.

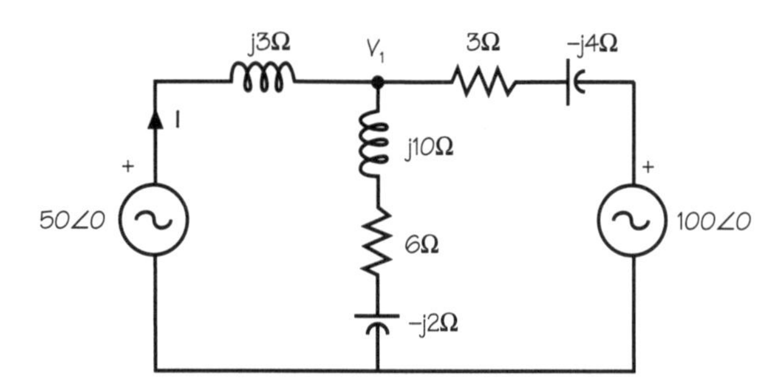

The node voltage equation at V1 is written as:

$$\mathbf{V}_1[1/(6 + j10 - j2) + 1/j3 + 1/(3 - j4)] - 50/j3 - 100/(3 - j4) = 0$$
$$\mathbf{V}_1 = (12 - j0.67)/(0.18 - j0.25) = 39\angle 51.05 = 24.52 + j30.33$$
$$\mathbf{I} = (50 - 39\angle 51.05)/j3 = 13.2\ \angle{-140}\text{A}.$$

(Nodal analysis)

6.

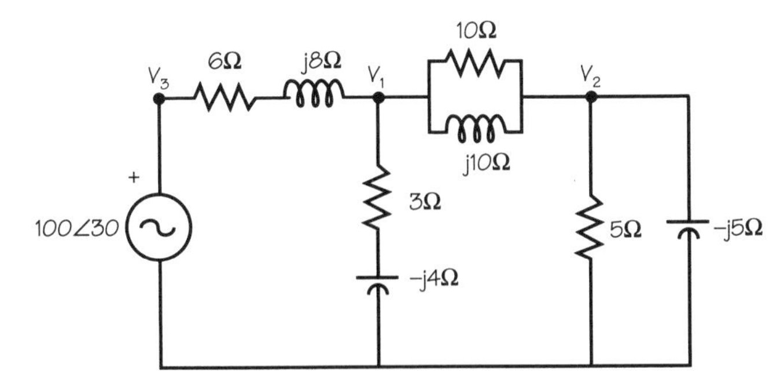

$\mathbf{V}_3 = 100\angle 30$ (is known).

Now, writing the node equations for $\mathbf{V}_1$ and $\mathbf{V}_2$:

$$\mathbf{V}_1[1/(6 + j8) + 1/(3 - j4) + 1/10 + 1/j10] - 100\angle 30/(6 + j8) - \mathbf{V}_2[1/10 + 1/j10] = 0 \quad (1)$$
$$-\mathbf{V}_1[1/10 + 1/j10] + \mathbf{V}_2[1/10 + 1/j10 + 1/5 + 1/-j5] = 0 \quad (2)$$

Equation (1) can be simplified to:

$$\mathbf{V}_1[0.04 - j.34] - \mathbf{V}_2\, 0.1414\angle -45 = 100\angle -23.13$$

Equation (2) can be simplified to yield: $\mathbf{V}_2 = 0.45\angle -63.43\mathbf{V}_1$

Inserting the above relationship in (1), we get:

$\mathbf{V}_1 = 349\angle 54.77$V.

$\mathbf{V}_2 = 157\angle -8.66$V. **(Nodal analysis)**

7.

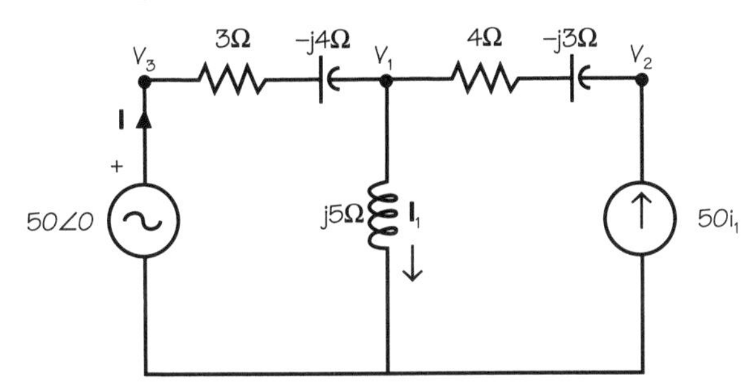

$\mathbf{V}_3 = 50\angle 0$ (is known)

Writing the node equations for $\mathbf{V}_1$ and $\mathbf{V}_2$:

$$\mathbf{V}_1[1/j5 + 1/(3 - j4) + 1/(4 - j3)] - 50(0/(3 - j4) - \mathbf{V}_2/(4 - j3) = 0 \quad (1)$$
$$-\mathbf{V}_1[1/(4 - j3)] + \mathbf{V}_2[1/(4 - j3)] - 50i = 0 \quad (2)$$

where $\mathbf{I}_1 = \mathbf{V}_1/j5$ (3)

Using (3), (2) is simplified to yield: $\mathbf{V}_2 = 49.4\angle -125.94\ \mathbf{V}_1$

Using the above relationship in (1), we obtain:

$\mathbf{V}_1 = 1\angle -36.17$ V

$\mathbf{V}_2 = 49.4\angle -162.1$ V

$$\mathbf{I} = \frac{50\angle 0 - 1\angle -36.17}{3 + j4} = 9.84\angle -52.5$$

(Nodal analysis)

8.

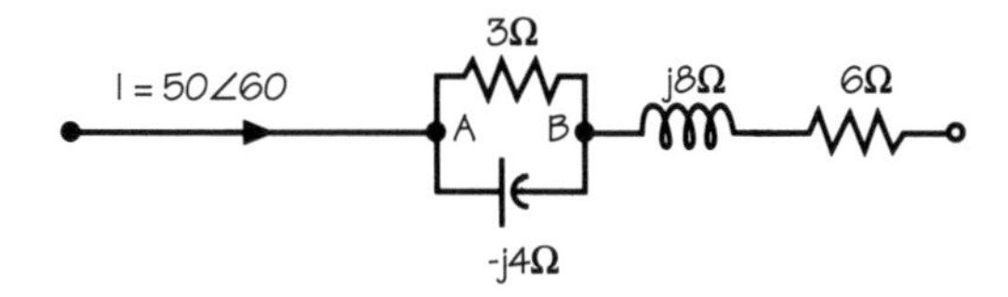

The voltage drop across the parallel combination of 3Ω and –j4Ω is

$$5\angle 60 + \frac{3 + -j4}{3 - j4} = 12\angle 23.13 \text{ V.}$$

(Nodal analysis)

9.

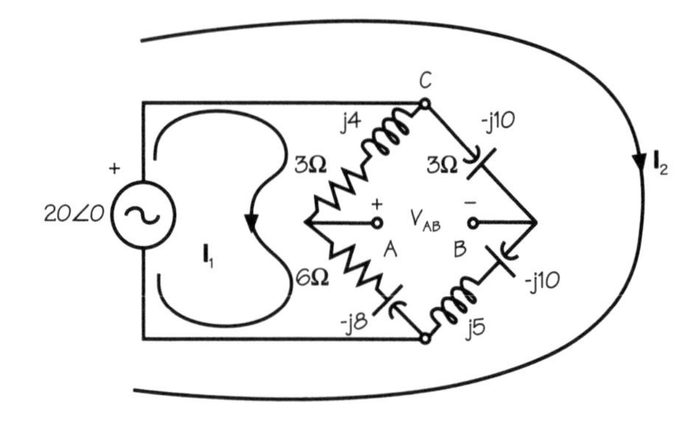

Using the method of mesh currents:

$20\angle 0 = (3 + j4 + 6 - j8)\mathbf{I}_1$
$20\angle 0 = (-j10 + j5 - j10)\mathbf{I}_2$. Solving for the currents give:
$\mathbf{I}_1 = 2.03\angle 23.96$A and
$\mathbf{I}_2 = 1.33\angle 90$A.

Now writing the KVL for the loop defined by ABC:

$$\mathbf{V}_{AB} + (3 + j4)\,\mathbf{I}_1 - (-j10)\mathbf{I}_2 = 0\text{: } \mathbf{V}_{AB} = V_{TH} = 14.69\angle -42.33\text{V}$$

Thevenin's impedance calculation:

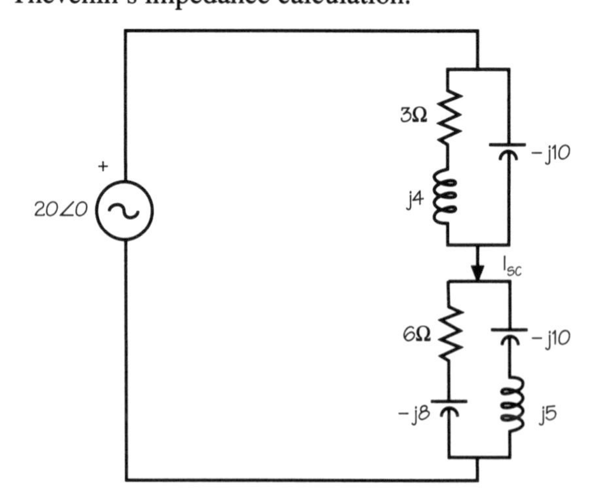

Taking the series/parallel combination of the circuit elements in the preceding circuit, the following circuit can be drawn:

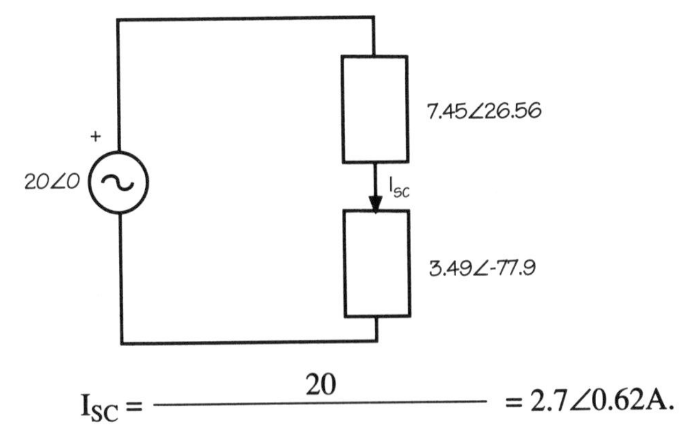

$$I_{SC} = \frac{20}{\qquad\qquad} = 2.7\angle 0.62\text{A}.$$

$\mathbf{Z}_{TH} = \mathbf{V}_{TH}/\mathbf{I}_{SC} = 5.428\angle 42.95\Omega.$

The resulting Thevenin's circuit is:

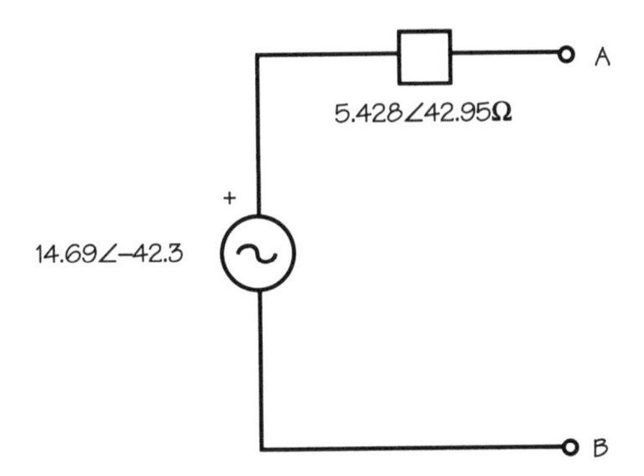

(Thevenin's theorem)

10.

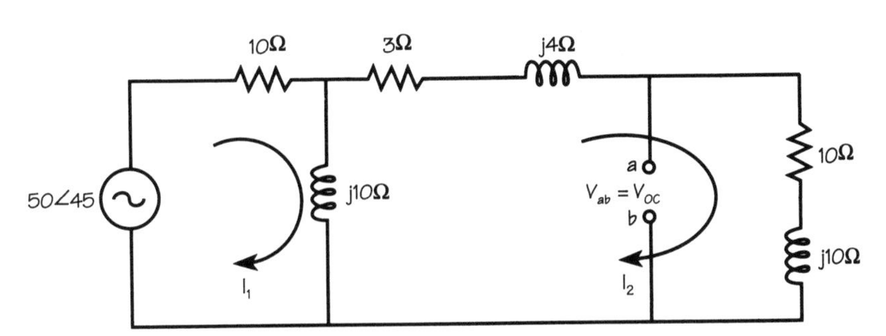

Using the method of mesh currents:

$$50\angle 45 = (10 + j10)\mathbf{I}_1 - j10\,\mathbf{I}_2 \qquad (1)$$
$$0 = -j10\,\mathbf{I}_1 + (13 + j\,24)\,\mathbf{I}_2 \qquad (2)$$

From (2): $\mathbf{I}_1 = \mathbf{I}_2(13 + j24)/j10$
substituting the above relation in (1):

$$50\angle 45 = -j(1 + j1)(13 + j24)\,\mathbf{I}_2 - j10\,\mathbf{I}_2.$$

Giving, $\mathbf{I}_2 = 1.3501\angle 43.45$ A.
Now, $\mathbf{V}_{AB} = \mathbf{I}_2(10 + j10) = 19.1\angle 88.45 = \mathbf{V}_{TH}$

Thevenin's impedance is calculated based upon the following circuit:

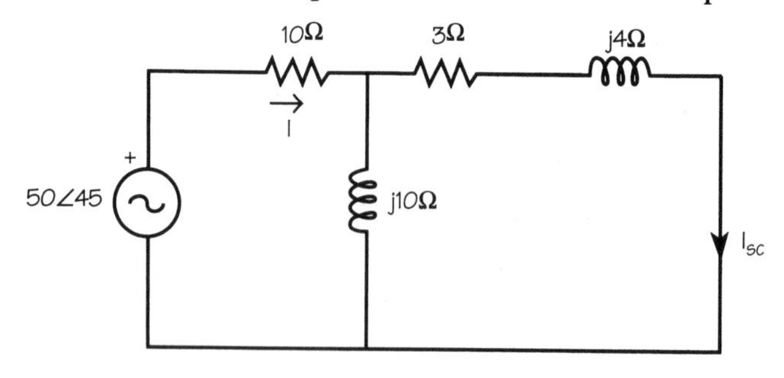

$$\mathbf{I} = \frac{50\angle 45}{10 + j10 \| (3 + j4)} = 4.2\angle 33.11 \text{ A.}$$

The short-circuit current $\mathbf{I}_{SC} = 4.2\angle 33.11 * j10/(3 + j14) = 2.933\angle 45.2$A.

$$\mathbf{Z}_{TH} = \mathbf{V}_{TH} / \mathbf{I}_{SC} = 19.1\angle 88.45 / 2.933\angle 45.2 = 6.5112\angle 43.25$$

The resulting Thevenin's equivalent circuit is:

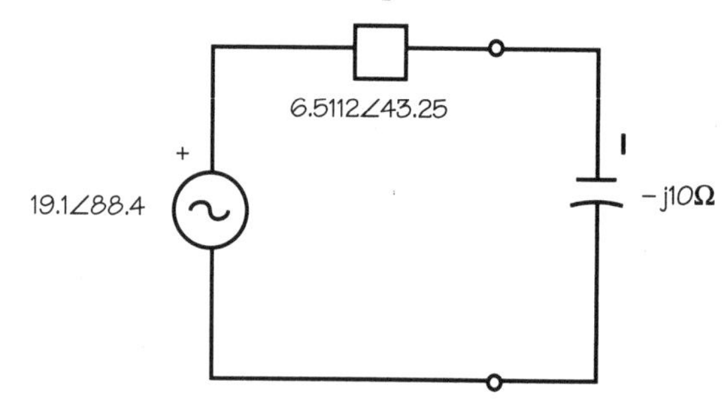

$$\mathbf{I} = 19.1\angle 88.45/(6.5112\angle 43.25 - j10) = 2.62\angle 137.9\text{A.}$$

(Thevenin's theorem)

11.

Z*TH = 5.428∠42.95

14.69∠–42.33

ZL = Z*TH

From problem 4.9, the Thevenin's equivalent circuit is:
For maximum power transfer, the load impedance is the complex conjugate of the line impedance:
$\mathbf{Z}_L = \mathbf{Z}_{TH}^* = 5.428\angle -42.95 = 3.973 - j3.698\Omega$.

(Maximum power transfer)

12.

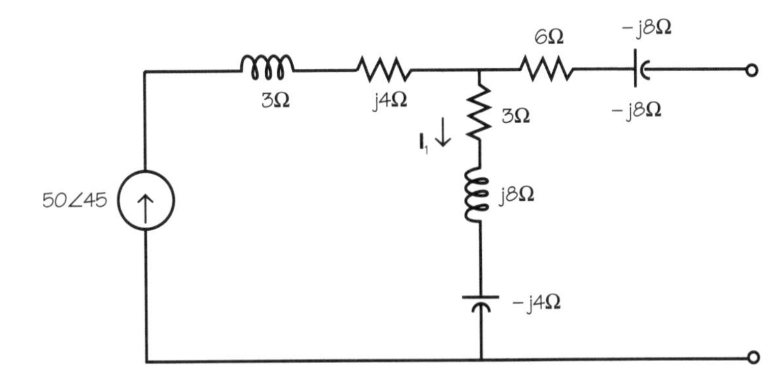

a) The component of current **I** through the RLC branch due to the voltage source alone $\mathbf{I}_1$ is determined. The current source is opened.

$\mathbf{I}_1 = 50\angle 45/(6{+}j8) = 5\angle{-8.13}$ A.

b) The voltage source is shorted and the current $\mathbf{I}_2$ due to the current source alone is calculated:

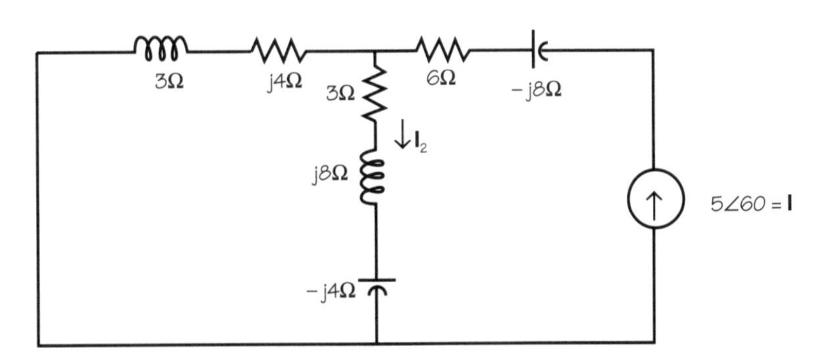

$\mathbf{I}_2 = 5\angle 60 * (3 + j4)/(6 + j8) = 2.5\angle 60$

c) The total current $\mathbf{I} = \mathbf{I}_1 + \mathbf{I}_2 = 5\angle{-8.13} + 2.5\angle 60 = 6.359\angle 13.25$A.

(Superposition)

Grade Yourself

Circle the numbers of the questions you missed, then fill in the total incorrect for each topic. If you answered more than three questions incorrectly, you need to focus on that topic. (If a topic has less than three questions and you had at least one wrong, we suggest you study that topic also. Read your textbook, a review book, or ask your teacher for help.)

Subject: AC Circuits

Topic	Question Numbers	Number Incorrect
RLC circuit complete solution (steady-state plus transient)	1	
Mesh current	2, 3, 4	
Nodal analysis	5, 6, 7, 8	
Thevenin's theorem	9, 10	
Maximum power transfer	11	
Superposition	12	

Operational Amplifiers

5

Brief Yourself

Op-amp circuits under dc excitation are introduced in this chapter. The op-amps are considered ideal, thereby justifying the use of virtual short between the inverting and non-inverting terminals. Use has been made of network theorems (chapters 2 and 4) to solve circuits containing one or more op-amps. Transient analysis questions involving op-amps are presented in this chapter. (Op-amp problem solution using Pspice is discussed in chapter 8.)

Test Yourself

1. For the circuit shown, determine v_0.

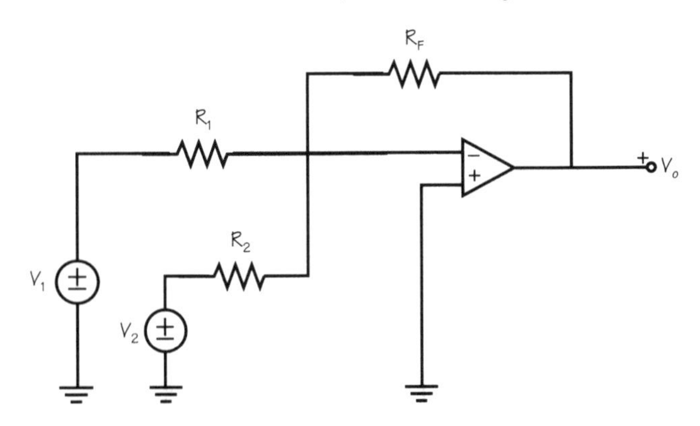

2. For the op-amp circuit shown, calculate v_0.

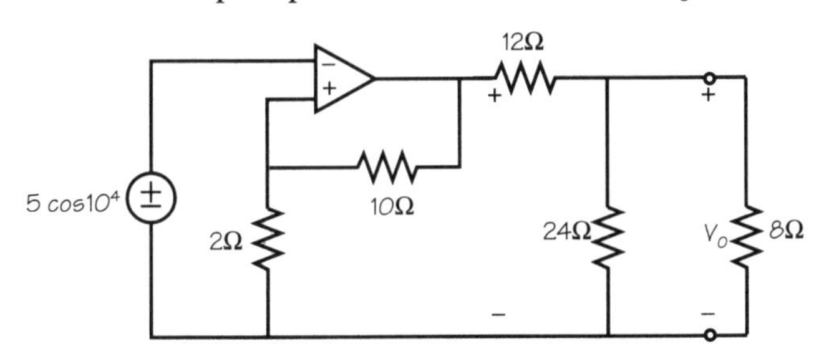

3. For the op-amp circuit shown, calculate the output voltage, assuming $R_F = R_1$.

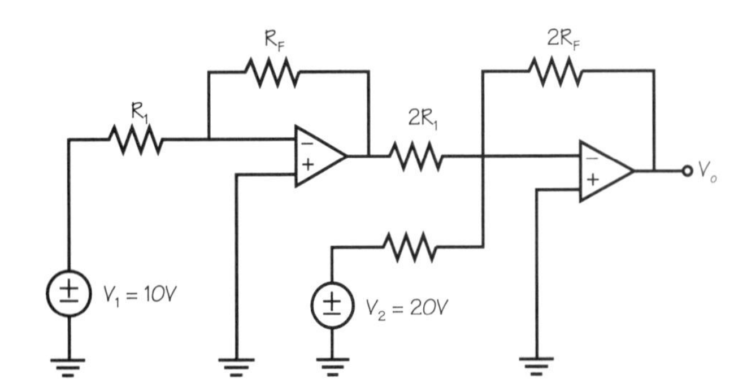

4. For the circuit shown in problem 5.2, calculate the current through the 8Ω resistance by first replacing the circuit by its Thevenin's equivalent.

5. Determine $v_0(t)$.

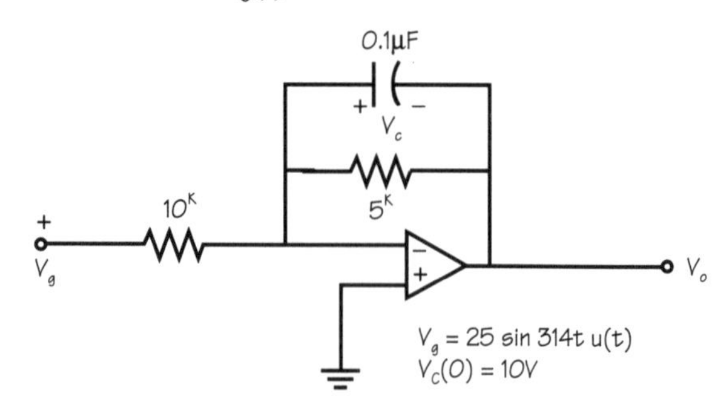

6. Find v(t) if $v_g(t) = 10e^{-10t}u(t)$. Consider the capacitor to be initially discharged.

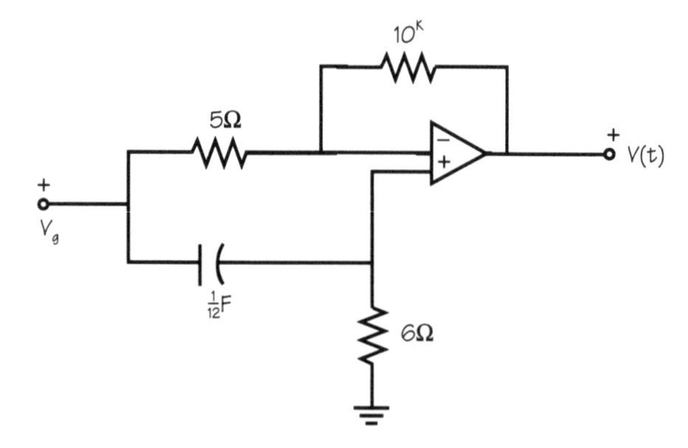

7. For initially discharged capacitors and $v_g(t)$ = 10 cos 10t, calculate $v_0(t)$.

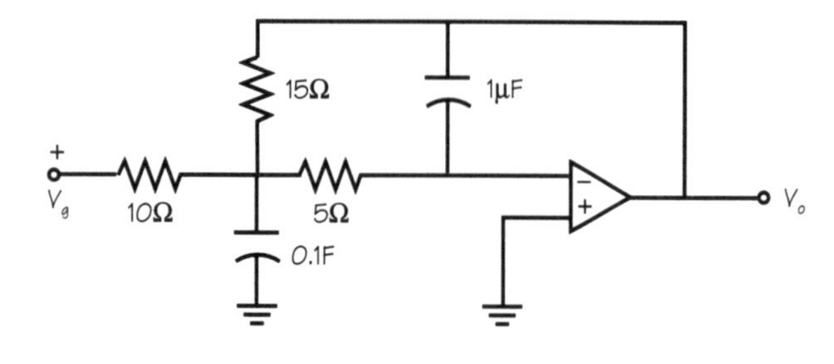

8. Comment on $v_0(t)$ for the circuit shown.

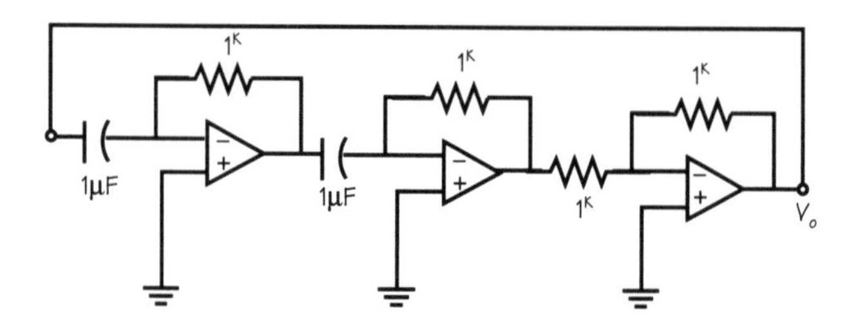

9. For the circuit shown, determine the gain $g = v_0/v_{in}$ at f = 1 Khz.

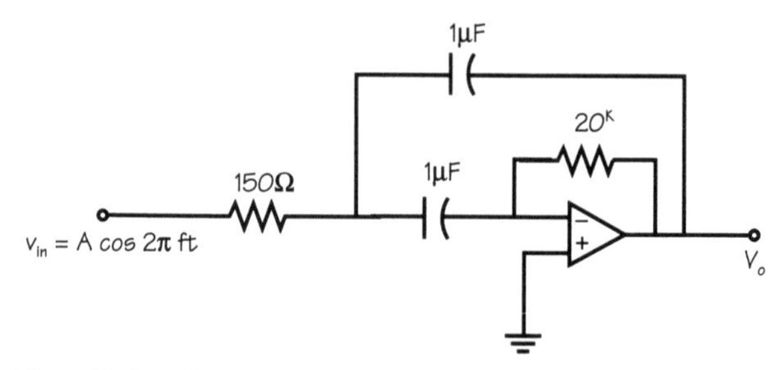

10. Solve for $v_0(t)$.

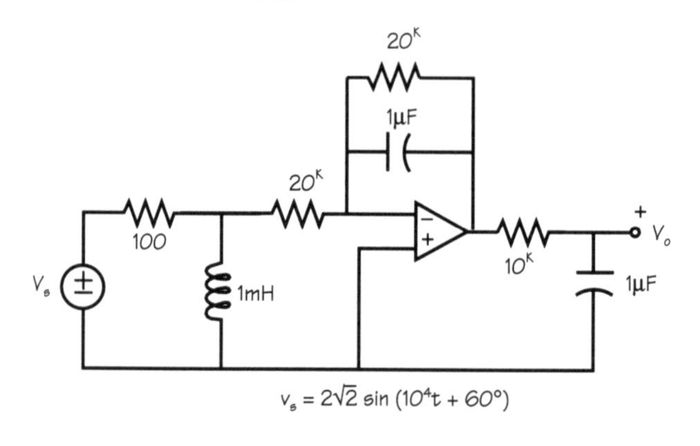

Check Yourself

1.

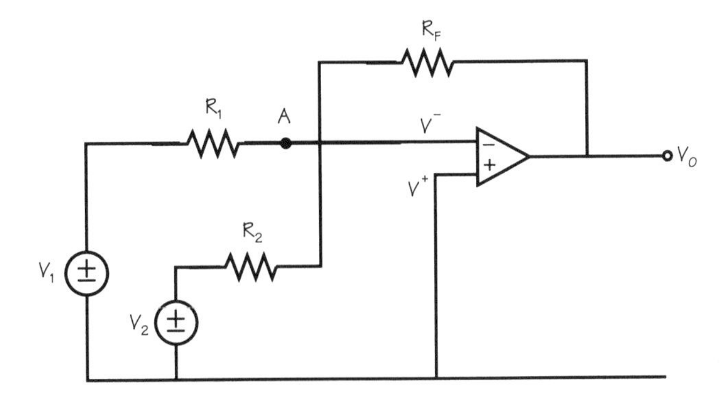

Considering ideal op-amp V− = V+ = 0V.
Applying KCL at node A: $V_2/R_L + V_1/R_1 = -V_0/R_F$

$$V_0 = -R_F(V_1/R_1 + V_2/R_2)$$

(Operational-amplifier circuits)

2.

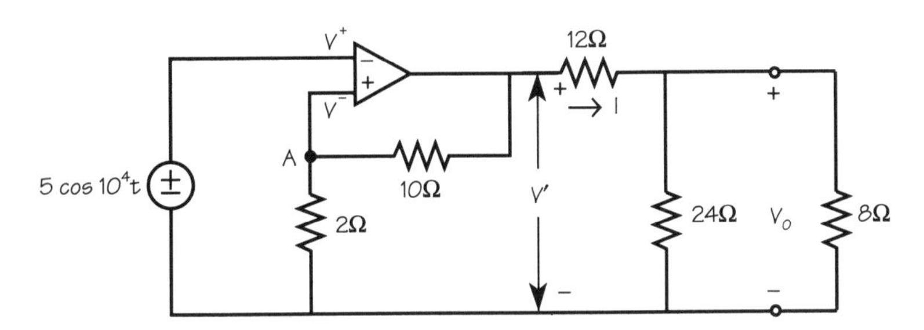

For an ideal op-amp $V+ = V- = 5\cos 10^4 t$.
Applying KCL at node A: $5\cos 10^4 t/2 + (5\cos 10^4 t - V')/10 = 0 \Rightarrow V' = 30\cos 10^4 t$.

Now, $V_0 = (24||8) * V'/(12 + 24||8) = 10\cos 10^4 t$

(Operational-amplifier circuits)

3.

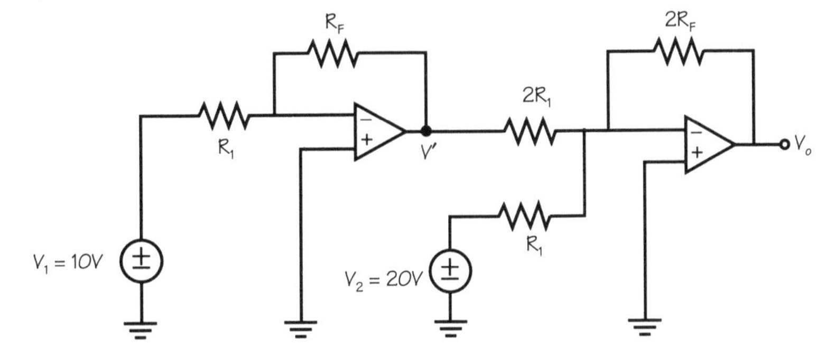

$V_A = V_B = 0$ (ideal op-amp).

Writing KCL at node A: $10/R_1 + V'/R_F = 0 \Rightarrow V' = -(R_F/R_1)\,V_1$.
Writing KCL at node B: $V'/(2R_1) + V_2/R_1 + V_0/(2R_F) = 0$

$$\Rightarrow V_0 = -(R_F/R_1)V' - (2R_F/R_1)V_2$$

Substituting for V′:

$$V_0 = -(R_F/R_1)[-(R_F/R_1)V_1 + 2V_2]$$

For $R_F = R_1$: $V_0 = -30V$.

(Operational-amplifier circuits)

4.

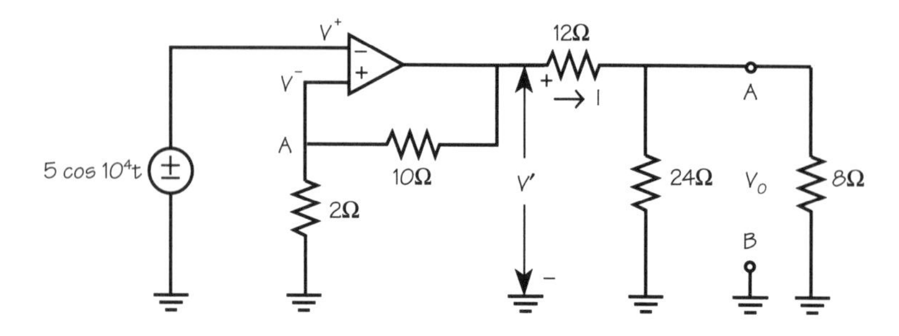

$V- = 5\cos 10^4t$.
V' is found to be $30\cos 10^4t$.

Therefore, $V_{AB} = V_{TH} = 24 * 30\cos 10^4t/(12 + 24) = 20\cos 10^4t$.
Thevenin's resistance is determined by using the following circuit:

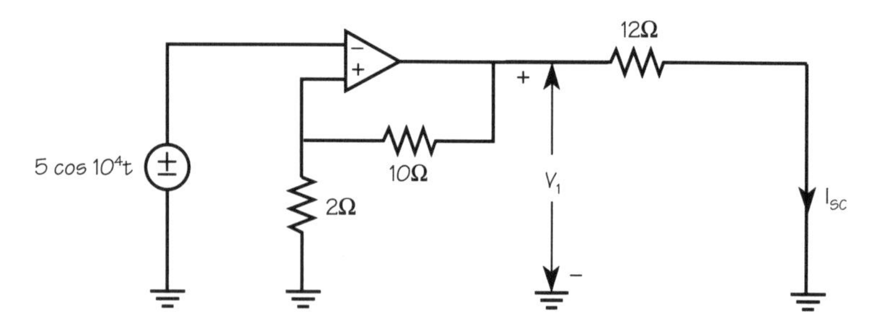

$V' = 30\cos 10^4t$. $I_{SC} = 30\cos 10^4t/12$
$R_{TH} = V_{TH}/I_{SC} = 8\Omega$.

The current through the 8Ω resistor is found by using the Thevenin's equivalent circuit:

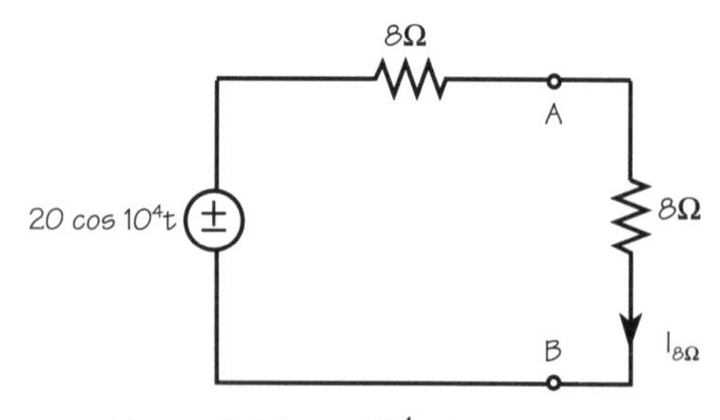

$I_{8\Omega} = 1.25\cos 10^4t$ A.

(Operational-amplifier circuits)

5.

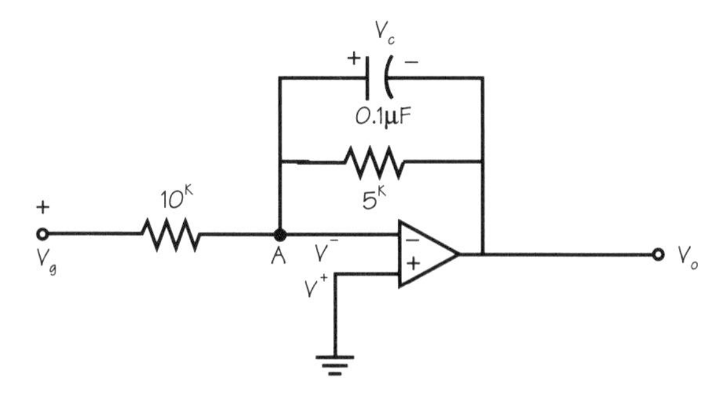

V– = 0V.

Applying KCL at node A:

$$\frac{v_g}{10^K} + \frac{v_0}{5^K} + C\frac{dv_0}{dt} = 0 \Rightarrow \frac{dv_0}{dt} + 2\times 10^3 v_0 = -10^3 v_g$$

Using integrating factors: $\int_0^t d(e^{2\times 10^3 t} v_0) = -10^3 \int_0^t v_g e^{2\times 10^3 t} dt$

The output potential in response to 2sin 314t can now be written as:

$$v_0(t)e^{2\times 10^3 t} - v_0(0) = -10^3 \mathrm{Re}\left[\int_0^t 2e^{(2\times 10^2 + j314)t} dt\right],$$ [here Re stands for the real part].

Upon solving the above equation, the output potential can be written as:

$$v_0(t) = 10 \exp(-2\times 10^{3t}) - 988\times 10^{-3} \sin(314t + 81°)\text{V}.$$

(Operational-amplifier circuits)

6.

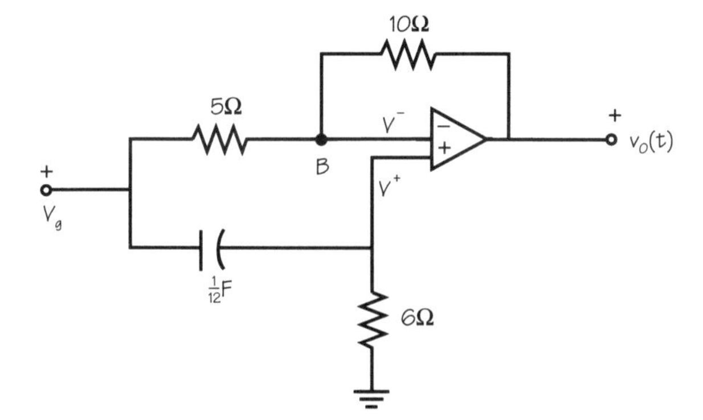

Part of the input circuitry connected to the non-inverting input is redrawn as:

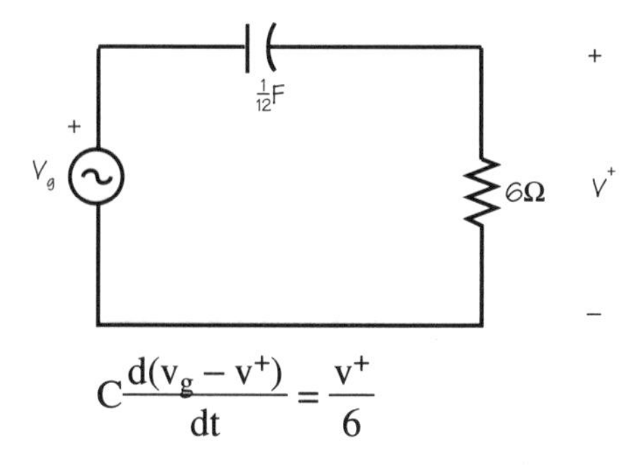

$$C\frac{d(v_g - v^+)}{dt} = \frac{v^+}{6}$$

That may be rewritten as:

$$\frac{dv^+}{dt} + 2v^+ = -100e^{-10t}.$$ Upon using the method of integrating factors, v^+ can be written as:

$$v^+ = 12.5\,(e^{-10t} - e^{-2t})$$

Now, for an ideal op-amp v+ = v–. Applying KCL at node B:

$$(v_g - v_-)/5 + (v\ - v_-)/10 = 0.$$ Therefore, $v(t) = 17.5\,e^{-10t} - 37.5e^{-2t}$

(Operational-amplifier circuits)

7.

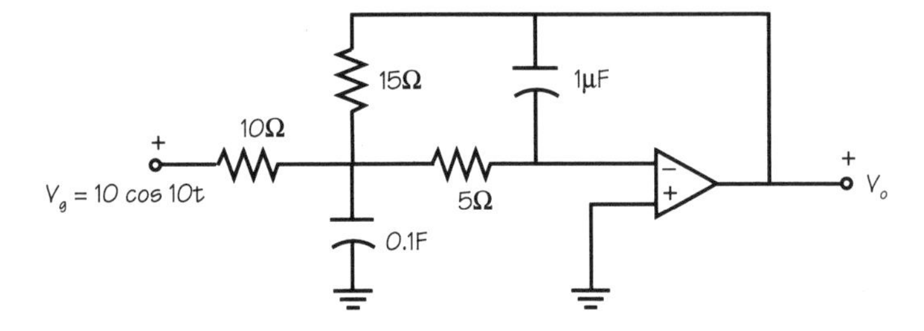

The forced response to the input voltage 10 cos 10t is solved using phasor technique, instead of using the method utilizing differential equations.

ω = 10rad/sec is the angular frequency. Using this information, the above circuit can be redrawn as:

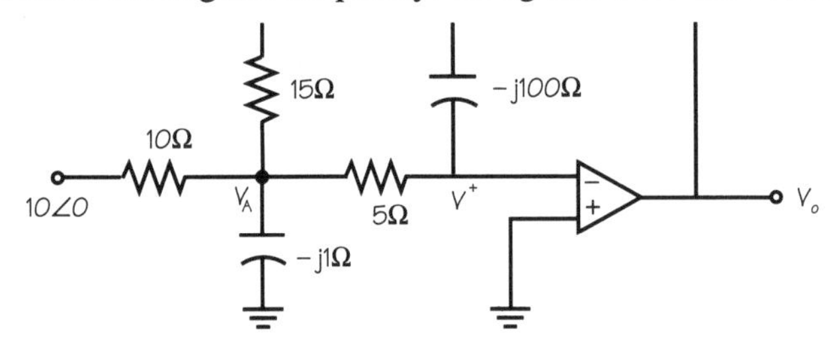

Writing KCL at V_A:

$$\mathbf{V}_A[1/10 + 1/{-j1} + 1/5 + 1/15] - \mathbf{V}_g/10 - \mathbf{V}_0/15 = 0 \qquad (1)$$

At node V–: $-\mathbf{V}_A/5 - \mathbf{V}_0/{-j100} = 0 \Rightarrow \mathbf{V}_A = -j0.05\,\mathbf{V}_0$. Using this relation in (1):

$$\mathbf{V}_A(0.367 + j1) = 0.1\,\mathbf{V}_g + \mathbf{V}_0/15.$$

Giving $\mathbf{V}_0 = 1\angle 0/(-0.0167 - j\,0.183) = 40.32\angle 132.33$ V.

Therefore, $v_0(t) = 40.32\cos(10t + 132.33)$V.

(Operational-amplifier circuits)

8.

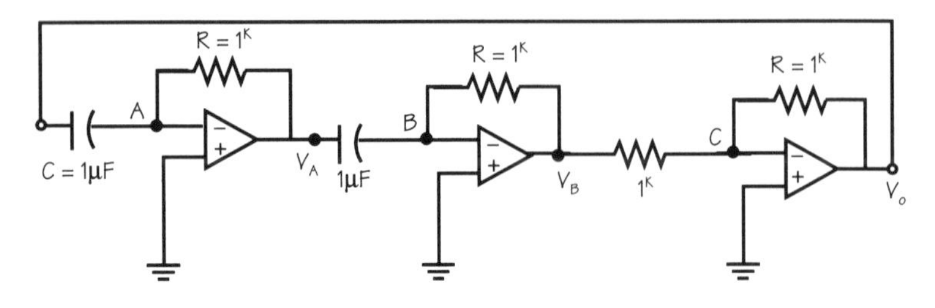

Applying KCL at A: $C\frac{dv_0}{dt} + \frac{v_A}{R} = 0 \qquad (1)$

Applying KCL at B: $C\frac{dv_A}{dt} + \frac{v_B}{R} = 0$ (2)

Applying KCL at C: $\frac{v_B}{R} + \frac{v_0}{R} = 0$ (3)

Using (3) and (2): $C\frac{dv_A}{dt} - \frac{v_0}{R} = 0$ (4)

Differentiating (1) with respect to t and using (4), we obtain:

$\frac{d^2v_0}{dt^2} + \frac{1}{R^2C^2}v_0 = \frac{d^2v_0}{dt_2} + \omega^2 v_0 = 0$. The above circuit will oscillate with a frequency of $f = 1/2\pi RC$.

(Operational-amplifier circuits)

9.

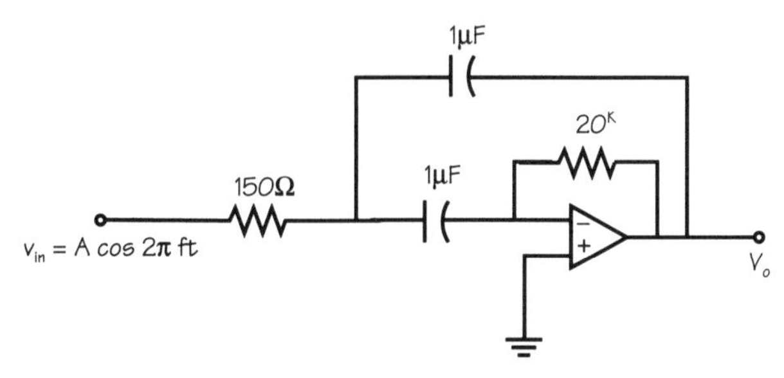

$\omega = 2\pi f = 6283$ rad/sec. (f = 1 kHz)

Capacitive reactance $X_c = 1/\omega C = 159\Omega$.

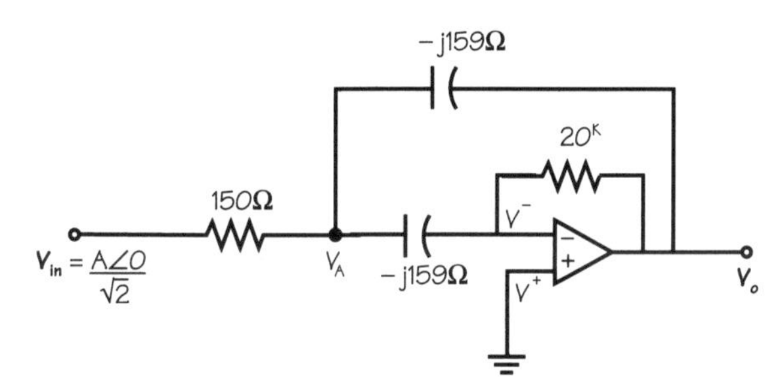

For an ideal op-amp: V+ = V– = 0.

Writing the node equation at $\mathbf{V}_A$:

$$\mathbf{V}_{in}/150 - \mathbf{V}_A[1/150 + 1/(-j159) + 1/(-j159)] - \mathbf{V}_0/-j159 = 0 \quad (1)$$

Node equation at V–: $-\mathbf{V}_A/(-j159) - \mathbf{V}_0/20^K = 0 \Rightarrow \mathbf{V}_A = j7.96 \times 10^{-3}\mathbf{V}_0$.
Eq. (1), can now be rewritten as:

$$\mathbf{V}_{in}/150 - j7.96 \times 10^{-3}[6.67 \times 10^{-3} + j12.6 \times 10^{-3}]\,\mathbf{V}_0 - j6.3 \times 10^{-3}\,\mathbf{V}_0 = 0$$

Therefore, $g = \mathbf{V}_0/\mathbf{V}_{in} = 1.05\angle -89.1$

(Operational-amplifier circuits)

10.

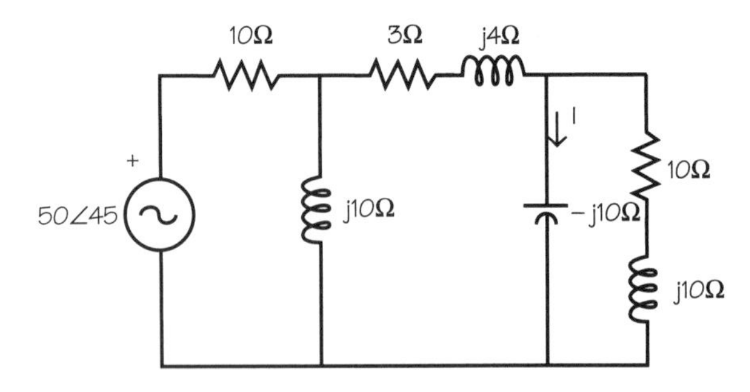

$\omega = 10{,}000$ rad/sec
Inductive reactance $X_L = \omega L = 10\Omega$.
Capacitive reactance $X_C = 1/\omega C = 100\Omega$.

The op-amp circuit is redrawn as:

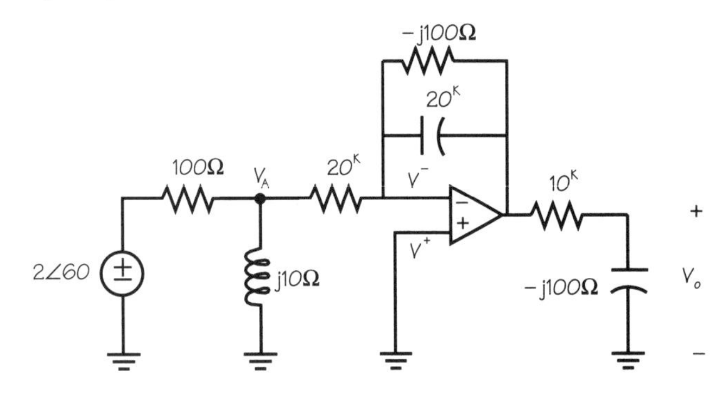

Applying KCL at VA: $\mathbf{V}_A[1/100 + 1/20^K + 1/j10] - 2\angle 60/100 = 0$
Giving: $\mathbf{V}_A = 199 \times 10^{-3}\angle 144.26$ V

Applying KCL at V−: $-\mathbf{V}_A/20K - \mathbf{V}_B[1/20K + 1/{-j100}] = 0$
Implying $\mathbf{V}_B = -995 \times 10^{-6}\angle 54.56$ V

Therefore, $V_0 = (j100)\ 995 \times 10^{-6}\angle 54.56\ /(-j100 + 10K) = 995 \times 10^{-6}\angle 145$ V.

$$v_0(t) = 995\sqrt{2} \times 10^{-6} \sin(10^4 t + 145°)$$

(Operational-amplifier circuits)

Grade Yourself

Circle the numbers of the questions you missed, then fill in the total incorrect for each topic. If you answered more than three questions incorrectly, you need to focus on that topic. (If a topic has less than three questions and you had at least one wrong, we suggest you study that topic also. Read your textbook, a review book, or ask your teacher for help.)

Subject: Operational Amplifiers

Topic	Question Numbers	Number Incorrect
Operational-amplifier circuits	1, 2, 3, 4, 5, 6, 7, 8, 9, 10	

Phasor Analysis

Brief Yourself

Phasor analysis greatly simplifies the analysis of steady-state ac circuits under sinusoidal excitation. For non-sinusoidal signals the concept of impedance does not hold, and the method of phasors cannot be used. Special techniques to represent non-sinusoidal signals in terms of a series of sinusoids (Fourier techniques) can then be applied and the concept of impedance can be revived. Network theorems, introduced in chapter 2, are perfectly valid for phasors and are extensively used in this chapter. It should be noted that phasor quantities retain the effective magnitude of current and voltage and their phase; however, the frequency information is lost.

Test Yourself

1. For the series circuit shown, calculate the voltage drop across the elements and draw the complete phasor diagram.

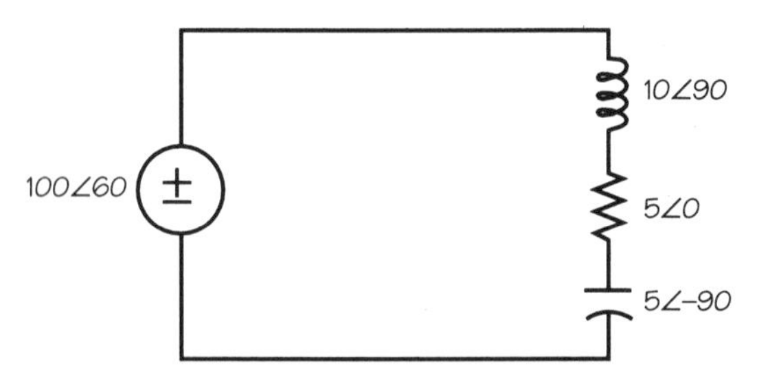

2. For the circuit shown, determine the equivalent impedance.

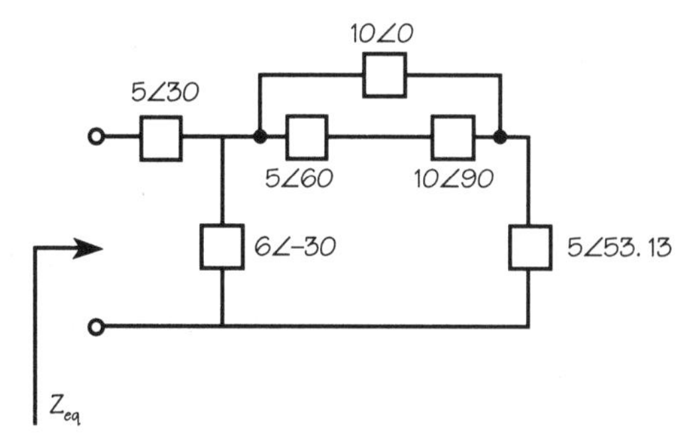

3. For the circuit shown, determine the equivalent impedance.

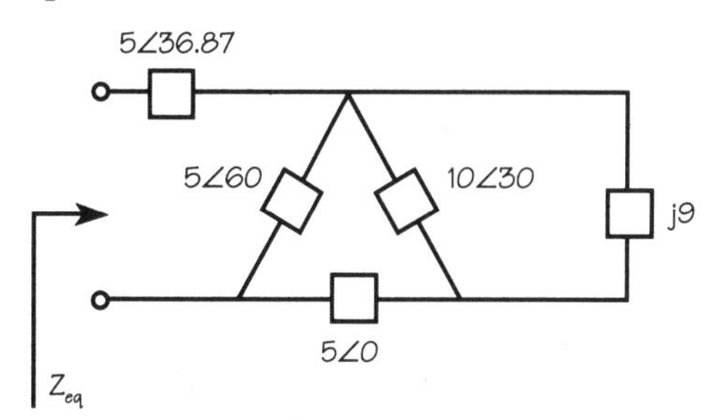

4. For the RC circuit, determine the total response.

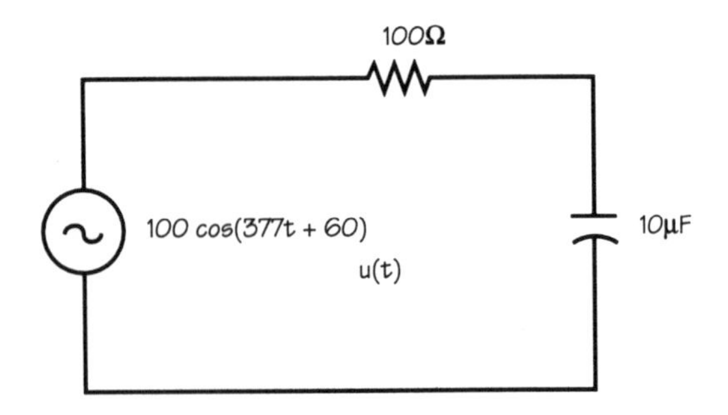

5. Using phasors, calculate the voltage drop across the load.

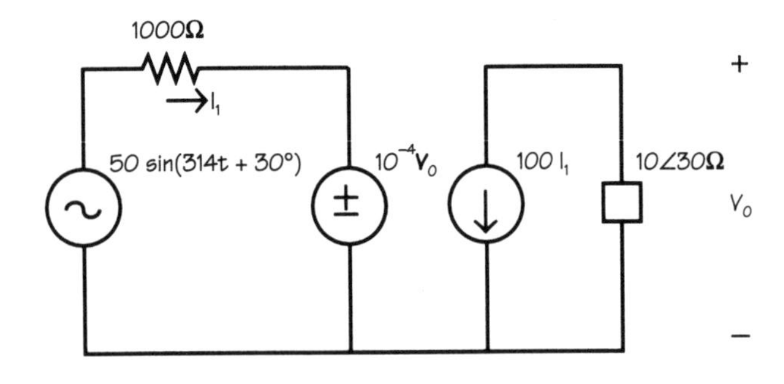

6. In the op-amp circuit shown, calculate the load current.

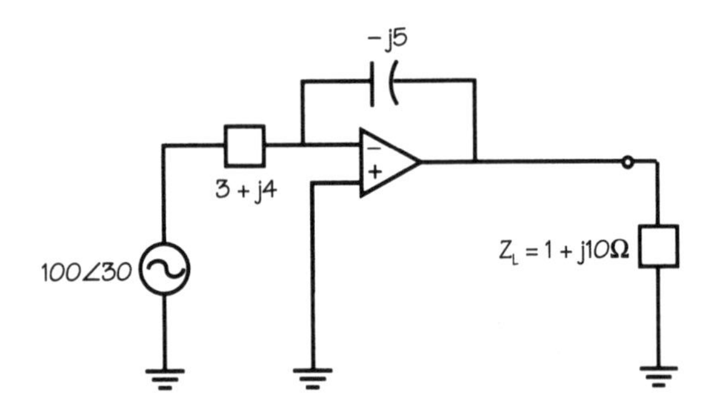

7. In the bridge circuit shown, determine V_{AB}.

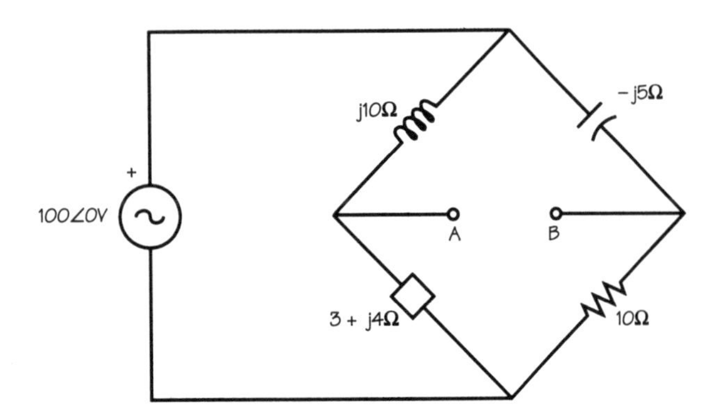

8. Draw the phasor diagram when the circuit is in reasonance. Comment on the total impedance.

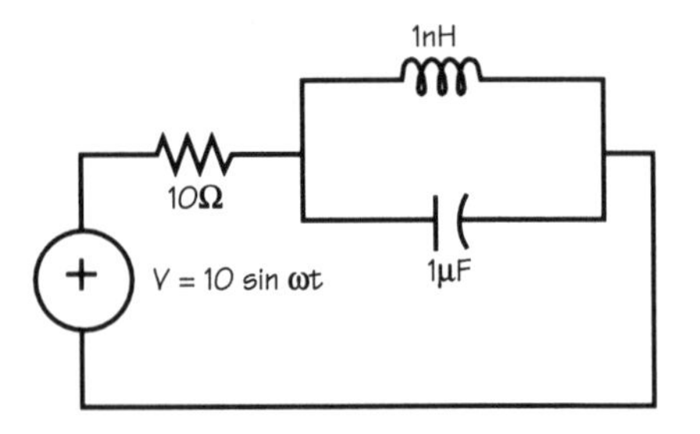

9. A FET with an inductive load is shown. Calculate the gain.

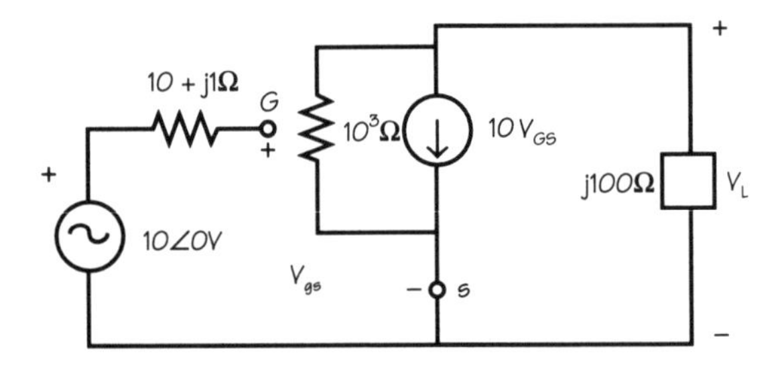

10. Draw the phasor diagram when the circuit is in resonance.

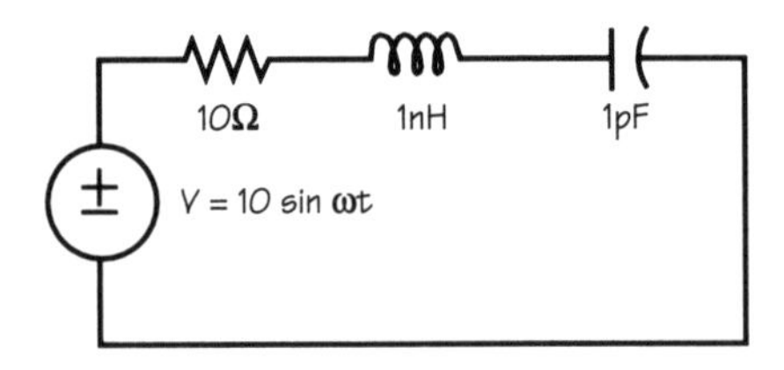

Check Yourself

1. 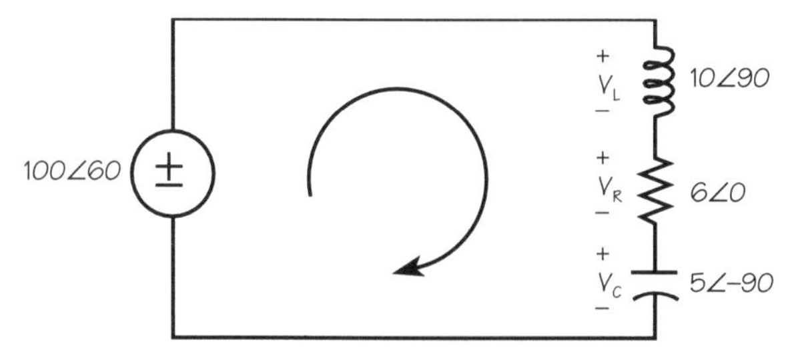

$$I = \frac{100\angle 60}{6 + j(10 - 5)} = \frac{100}{6 + j5}\angle 60 - 39.8$$

$I \Rightarrow I = 12.8\angle 20.2$

$V_L = 12.8\angle 20.2 * 10\angle 90 = 128\angle 110.2$ V

$V_C = 5\angle\text{-}90 * 12.8\angle 20.2 = 64\angle -69.8$ V

$V_R = 6 * 12.8\angle 20.2 = 76.8\angle 20.2$ V

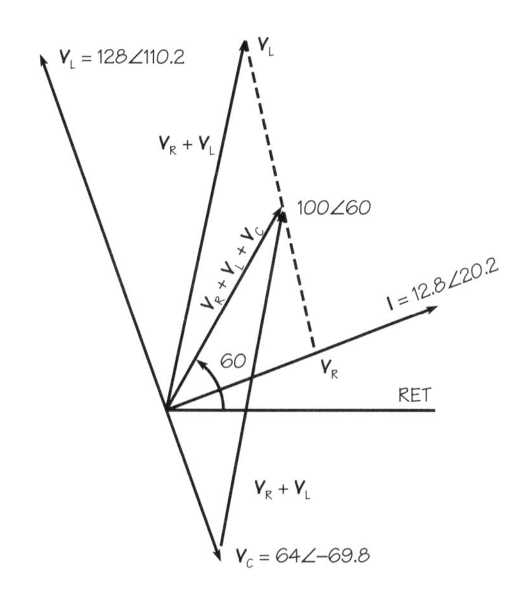

(Series RLC)

2.

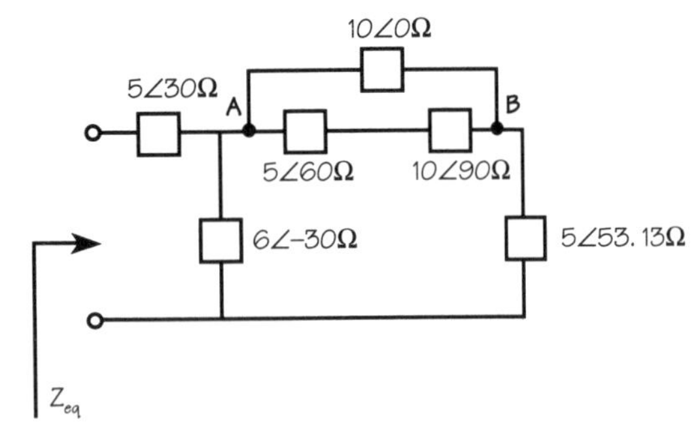

$$Z_{AB} = (5\angle 60 + 10\angle 90) \| 10\angle 0$$

$$5\angle 60 + 10\angle 90 = 2.5 + j14.33 = 14.55\angle 80.1$$

$$Z_{AB} = \frac{14.5\angle 80.1}{12.5 + j14.33} = 7.65\angle 31.2\Omega$$

$$Z_{eq} = 5\angle 30 + 6\angle -30 \| (7.65\angle 31.2 + 5\angle 53.13)$$

$$= 5\angle 30 + 6\angle -30 \| 12.43\angle 39.84$$

$$= 5\angle 30 + \frac{74.58\angle 9.84}{15.55\angle 18.6}$$

$$\Rightarrow Z_{eq} = 5\angle 30 + 4.796\angle -8.76$$

$$Z_{eq} = 9.24\angle 11.04\Omega$$

(Series-parallel)

3.

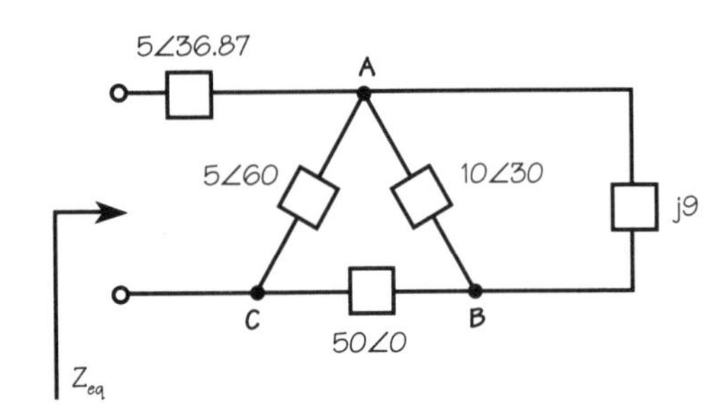

The Δ connection can be converted to a Y(WYF); and Z_{eq} can be evalated. However, the present circuit is tractable only using series and parallel combination of resistors.

$$Z_{AB} = 10\angle 30 \| j9 = 5.467\angle 61.74$$

$$Z_{AC} = 5\angle 60 \| (5.467\angle 61.74 + 50) = 5.176\angle 56.5$$

$$Z_{eq} = 5\angle 36.867 + 5.176\angle 56.5 = 10\angle 46.89\Omega$$

(Δ-Y)

4.

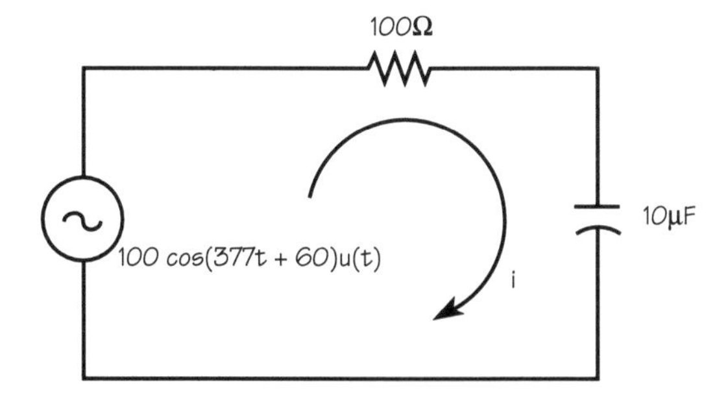

The total response will consist of the transient and the steady-state response.

$$i(t) = i_{TR}(t) + i_{SS}(t)$$

$i_{TR}(t) = Ae^{-t/RC} = Ae^{-10^3 t}$, where A is an arbitrary constant.

Using phasor notation, the above circuit can be redrawn to determine the steady-state current:

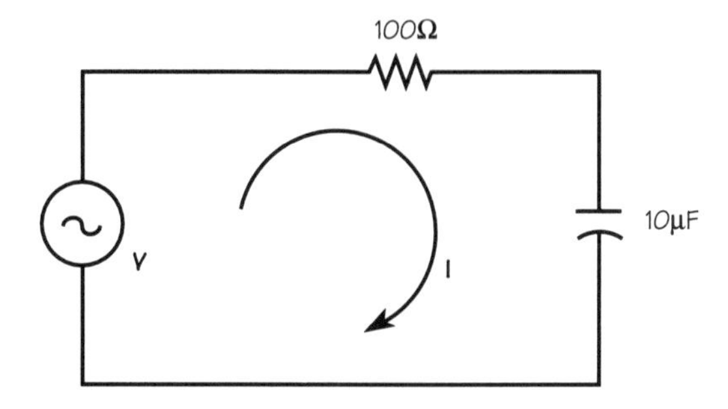

$$\mathbf{V} = \frac{100}{\sqrt{2}}\angle 60\text{V}$$

$$X_C = -j\frac{1}{377 * 10\mu F} = -j265.25\Omega$$

$$\mathbf{I} = \frac{\frac{100}{\sqrt{2}}\angle 60}{100 - j265.25} = 0.249\angle 129.34\text{A}.$$

$$i_{SS(t)} = \sqrt{2}|\mathbf{I}|\cos(377t + 129.34°)$$

$$= 0.353\cos(377t + 129.34°)\text{ A}.$$

Now: $i(t) = Ae^{-10^3 t} + 0.353\cos(377t + 129.34°)$

Now: $V_c(0^+) = V_c(0^-) = 0$ [V_c is the voltage drop across the cap.]

$\Rightarrow$ at $t = 0$, $V_R = Ri(t) = 100\cos(60°)$

$\Rightarrow \cos 60 = A + 0.353\cos(129.34°)$

$\Rightarrow A = 1.134$

$\Rightarrow i(t) = 0.353\cos(377t + 129.34°) + 1.134e^{-1000t}\text{A}.$

(Total response)

5. This circuit represents a small signal BJT amplifier with an inductive load.

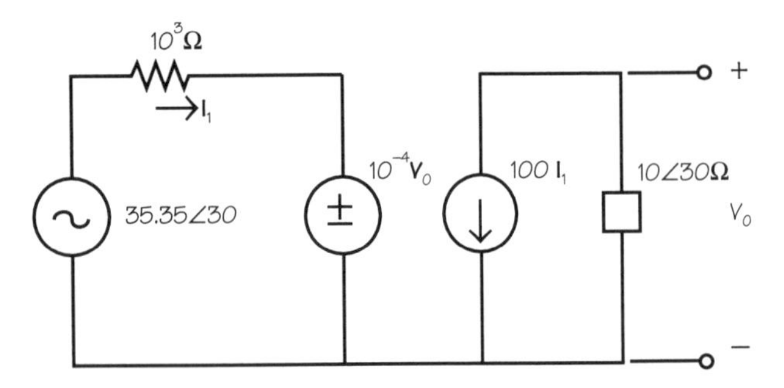

At the input:

$$\mathbf{I}_1 = \frac{35.35\angle 30 - 10^{-4}v_0}{10^3}$$

At the output:

$$\mathbf{V}_0 = -10\angle 30 * 100\mathbf{I}_1 = -10^3\angle 30\frac{35.35\angle 30 - 10^{-4}\mathbf{V}_0}{10^3}$$

$$\Rightarrow \mathbf{V}_0 = -35.35\angle 60\text{V}.$$

V = 35.35∠30

30°

120°

$V_0 = 35.35\angle -120$

(Simple BJT circuit)

6.

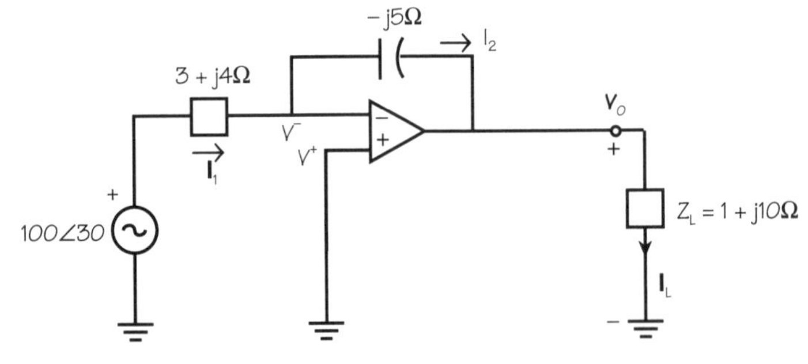

Assuming ideal op-amps $V^- = V^+ = 0$.
From the circuit $I_1 = I_2$

$$\Rightarrow \frac{100\angle 30 - 0}{3 + j4} = \frac{-\mathbf{V}_0}{-j5}$$

$$\Rightarrow \mathbf{V}_0 = j5\frac{100\angle 30}{5\angle 53.13} = 100\angle 66.87 \text{ Volt}$$

$$\Rightarrow \mathbf{I}_L = \frac{100\angle 66.87}{1 + j10} = 9.95\angle -17.42 \text{ Amp}$$

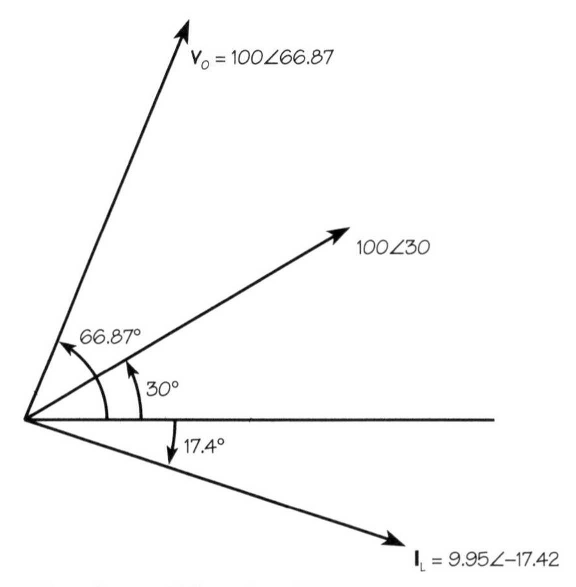

(Operational-amplifier circuit)

7.

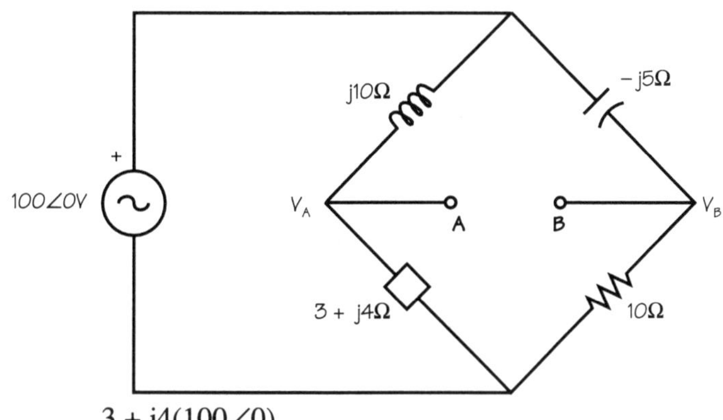

$$V_A = \frac{3 + j4(100\angle 0)}{3 + j4 + j10} = 34.92\angle -24.77$$

$$V_B\ 100\angle 0 \frac{10}{10 - j5} = 89.44\angle 26.56$$

$$V_{AB} = V_A - V_B = 34.92\ \angle -24.44 - 89.44\angle 26.56$$

$$V_{AB} = 72.9\angle -131.47$$

(Bridge circuit)

8. The LC combination constitutes a parallel resonance (tank) circuit. The resonant frequency is:

$$\omega_0 = \frac{1}{\sqrt{LC}} \text{ rad/sec}$$

$$\Rightarrow f = \frac{1}{2\pi \sqrt{LC}} \text{ Hz}$$

$$f = 5.033 \times 10^6 \text{ Hz} = 5.033 \text{ MHz}$$

At resonance:

$$jX_L = j\omega L = j(2\pi f)L = j31.623 \times 10^{-3}\Omega$$

$$-jX_C = -j\frac{1}{\omega C} = -j31.623 \times 10^{-3}\Omega$$

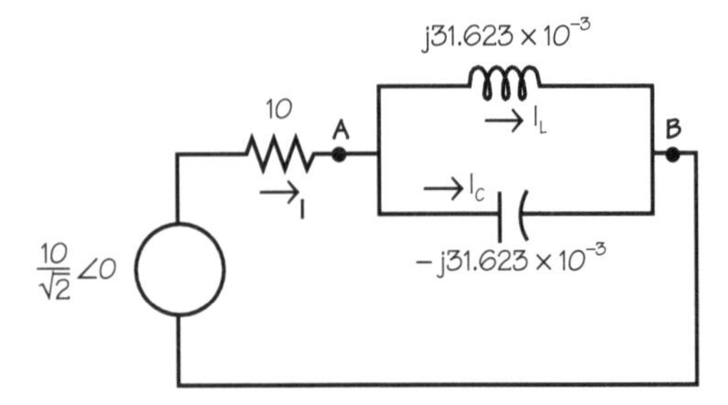

$Z_{AB} = jX_L \| (-jX_C)$ is infinitely large. A tank (ideal) circuit, therefore, at resonance behaves like an open circuit.

$$\mathbf{I}_L = \frac{\frac{10}{\sqrt{2}}\angle 0}{j31.623 \times 10^{-3}} = -j141.4 \times 10^{-3}$$

$$\mathbf{I}_C = \frac{\frac{10}{\sqrt{2}}\angle 0}{j31.623 \times 10^{-3}} = +j141.4 \times 10^{-3}$$

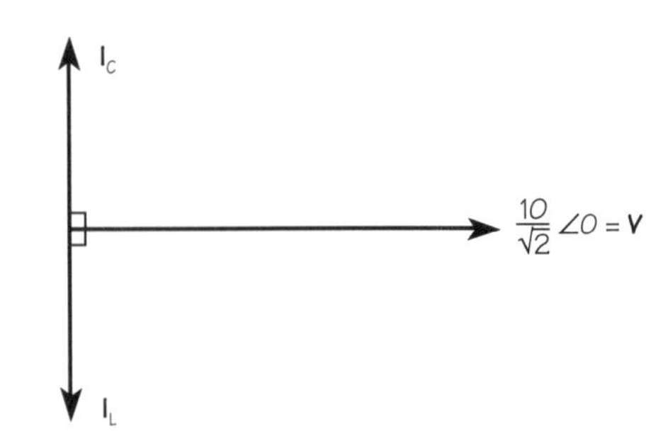

$$\Rightarrow \mathbf{I} = \mathbf{I}_L + \mathbf{I}_C = 0 \quad \textbf{(Parallel resonance)}$$

9.

At the input: $\mathbf{V}_{GS} = 10\angle 0$

$$\Rightarrow \mathbf{I}_L = \frac{\mu V_{GS}}{10^3 + j100} \text{ where, } \mu = 10 * 10^3 = 10^4 \text{ [NOTE: } \mu = g_m\, rd = 10 * 10^3\text{]}$$

$$\Rightarrow \mathbf{I}_L = 99.5\angle -5.7$$

$$\Rightarrow \mathbf{V}_L = j100 * \mathbf{I}_L = 9.95 \times 10^3 \angle 84.3\text{V}$$

Gain:

$$\left|\frac{\mathbf{V}_L}{10}\right| = 9.95 \times 10^2. \ \textbf{(FET circuit)}$$

10.

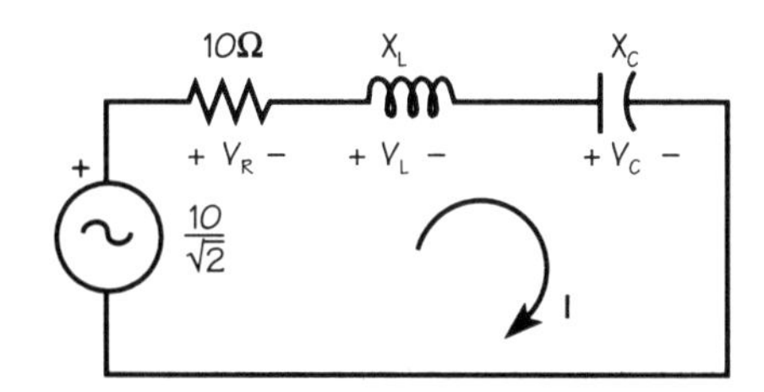

The circuit shown is a series-resonant circuit. The resonant frequency is determined by a similar relationship:

$$\omega_0 = \frac{1}{\sqrt{LC}}$$

$$\Rightarrow f = \frac{1}{2\pi\sqrt{LC}} = 5.023 \times 10^6 \text{ Hz} = 5.023 \text{ MHz}$$

$$X_L = j\omega L = j31.623 \times 10^3\Omega$$

$$X_C = j\frac{1}{\omega C} = -j31.623 \times 10^3\Omega.$$

$$\mathbf{I} = \frac{10/\sqrt{2}}{10 + j(31.623 - 31.623) \times 10^3} = 0.707\text{A}$$

Note: In series resonance, the circuit is purely resistive and shows the minimum impedance.

$$\mathbf{V}_R = \mathbf{I}R = 7.07 \text{ V}$$

$$\mathbf{V}_C = X_C\mathbf{I} = 2.23 \times 10^4\angle{-90} \text{ V}$$

$$\mathbf{V}_L = X_L\mathbf{I} = 2.23 \times 10^4\angle 90 \text{ V}$$

Note: At resonance, the voltage drop across the reactive elements can be many times greater than the supply voltage.

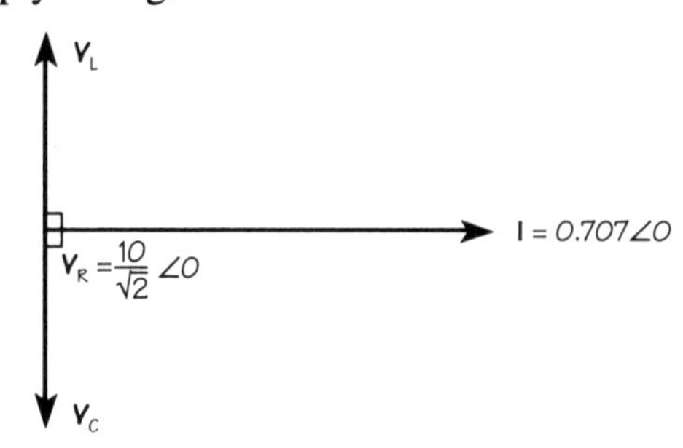

(Series resonance)

Grade Yourself

Circle the numbers of the questions you missed, then fill in the total incorrect for each topic. If you answered more than three questions incorrectly, you need to focus on that topic. (If a topic has less than three questions and you had at least one wrong, we suggest you study that topic also. Read your textbook, a review book, or ask your teacher for help.)

Subject: Phasor Analysis

Topic	Question Numbers	Number Incorrect
Series RLC	1	
Series-parallel	2	
Δ-Y	3	
Total response	4	
Simple BJT circuit	5	
Operational-amplifier circuit	6	
Bridge circuit	7	
Parallel resonance	8	
FET circuit	9	
Series resonance	10	

Power and Power Factor

Brief Yourself

The average power (more than one cycle) absorbed/dissipated by a purely capacitive or inductive circuit is zero. On the other hand, average power absorbed/dissipated by a resistor is finite for finite current and voltage. Unlike dc circuits, steady-state power calculation in ac circuits is complicated by the presence of reactive elements. (The reactive elements are either shorted or open in dc case.) In general, in ac circuits power is complex. The real part of the complex power happens to be the useful power, expressed in watts. The imaginary part of the complex power is the reactive power, and is expressed in VAR (volt ampere reactive). Our goal is to extract as much real power as possible from complex power, and that necessitates that the phase angle in between the current and voltage (required to calculate complex power) be as small as possible; that brings us to the definition of power factor angle. Power factor is defined as the ratio of real power to the magnitude of complex power.

Test Yourself

1. Using the method of phasors, calculate the real and reactive power supplied by the source.

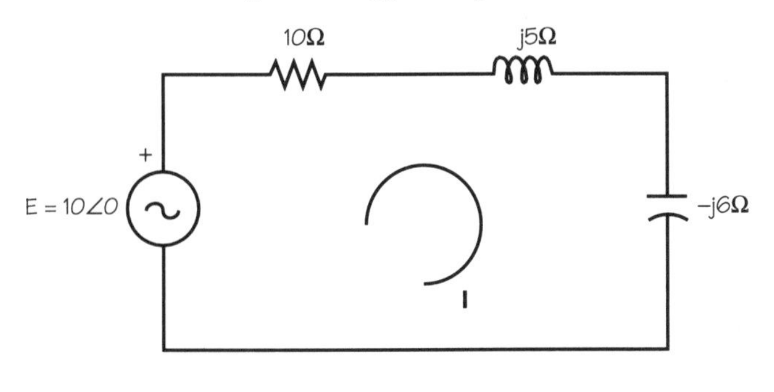

2. Calculate the apparent (complex), real, and reactive power absorbed by the load impedance Z_L.

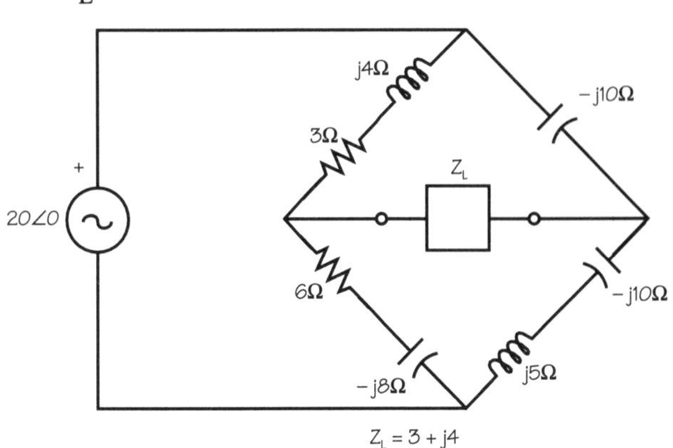

3. For the circuit shown, calculate the power factor angle for the two independent sources.

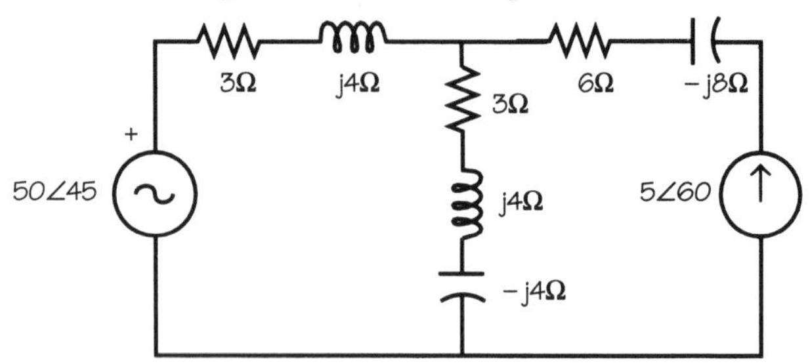

4. Construct a series circuit having **I** = 10∠45 A with a positive reactive power of 120 vars (assume supply voltage as reference).

5. In the parallel circuit shown, determine the reactive power absorbed by each branch. What is the power factor?

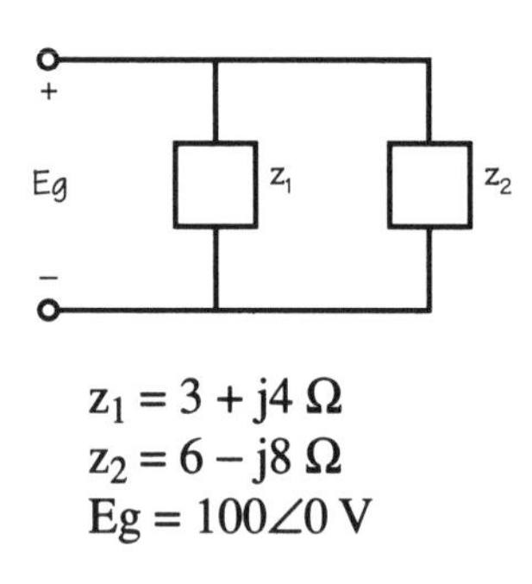

$z_1 = 3 + j4\ \Omega$
$z_2 = 6 - j8\ \Omega$
$Eg = 100\angle 0$ V

6. In the circuit shown in problem 7.2, determine z_L that will improve the overall power factor to 0.95 (lag).

7. The following loads are connected in parallel. Calculate the complex power and draw the complete power triangle.
   a) 50W at pf = 0.8 (lag)
   b) 100W at pf = 0.9 (lead)
   c) 200 VA, 100 var (capacitive)

8. Determine the voltage of the parallel capacitor to change the overall pf to: a) 0.9 lagging; and b) 0.8 lagging.
   z_L: 100VA at 0.6 (lag)

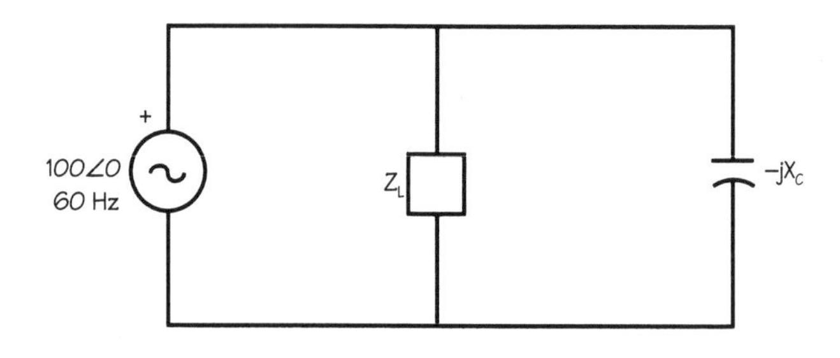

Check Yourself

1.

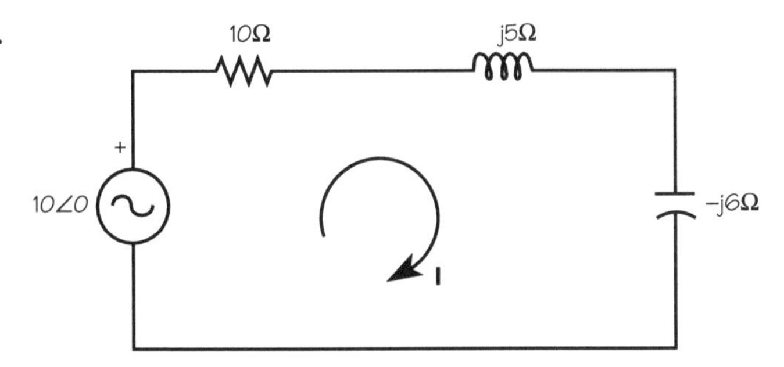

$$\mathbf{I} = \frac{10\angle 0}{10 + j5 - j6} = \frac{10\angle 0}{10.05\angle -5.7} = 09.95\angle 5.7$$

The phasor diagram can be drawn as:

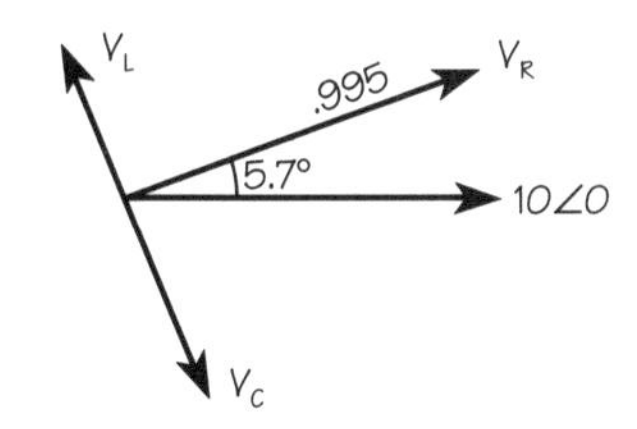

Real power supplied by the source.

$$P = |\mathbf{V}|\,|\mathbf{I}| \cos d$$
$$= 10 * 0.995 \cos (5.7)$$
$$P = 9.9 \text{ watt}$$

Reactive power, Q:

$$Q = |\mathbf{V}|\,|\mathbf{I}| \sin (\theta_V - \theta_I)$$
$$= 10 * 0.995 \sin(0 - 5.7)$$
$$Q = -0.9882 \text{ var}$$

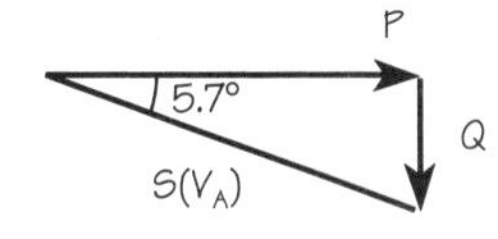

The power triangle:

$$VA = \sqrt{P^2 + Q^2} = |\mathbf{V}| * |\mathbf{I}|$$

The apparent power:

$$S = \sqrt{P^2 + Q^2} = |\mathbf{V}| * |\mathbf{I}| = 9.95 \text{ VA}$$

(Real and reactive power calculation)

2. Using Thevenin's theorem, the circuit can be redrawn as [see problem 4.9]:

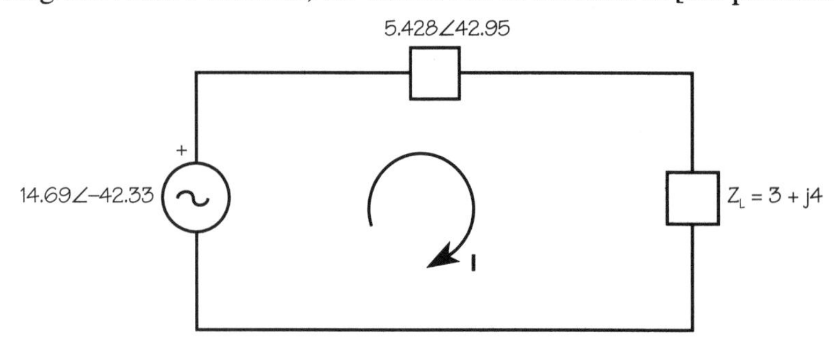

$$\mathbf{I} = \frac{14.69\angle{-42.33}}{5.428\angle 42.95 + 5\angle 53.13} = \frac{14.69\angle{-42.33}}{10.39\angle 47.83} \Rightarrow \mathbf{I} = 1.4143\angle{-90.16}\text{ A}$$

$\mathbf{V}_L = \mathbf{I} * Z_L = 1.4143 * 5\angle{-90.16} + 53.13 = 7.07\angle{-37}$

$P_L = |\mathbf{V}_L|\,|\mathbf{I}_L| \cos(\theta_V - \theta_I) = 7.07 * 1.4143 \cos(-37 + 90.16)$

$P_L = 6\text{W}$

$Q_L = 7.07 * 1.4143 \sin(-37 + 90.16) = 8$ vars

$|S| = |\mathbf{V}_L|\,|\mathbf{I}_L| = 10$ VA

Alternative Method:

$P_L = |\mathbf{I}|^2 R_L = (1.4143)^2 * 3 = 6\text{W}$

$Q_L = |\mathbf{I}|^2 X_L = (1.4143)^2 * 4 = 8$ vars

$S = \sqrt{P_L^2 + Q_L^2} = 10$ VA **(Real and reactive power calculation)**

3. From problem 4.12, we obtain:

$\mathbf{I}_{50V} = 5\angle{-8.13} - (5\angle 60 - 2.5\angle 60)$

$= 3.7 - j2.87 = 4.68\angle{-37.82}\text{A}$

To calculate the voltage drop across the current source, let us consider the method of node voltages.

$V_1 = 50\angle 45$

$$\frac{-50\angle 45}{3 + j4} + V_2\left[\frac{1}{3 + j4} + \frac{1}{3 + j4} + \frac{1}{6 - j8}\right] - \frac{V_3}{6 - j8} = 0$$

$$-V_2\left[\frac{1}{6 - j8}\right] + V_3\left[\frac{1}{6 - j8}\right] - 5\angle 60 = 0$$

V_1 V_2 V_3
3Ω j4Ω 3Ω 6Ω –j8Ω
50∠45 j8Ω 5∠60
I_{50V} –j4Ω

or: $-10\angle{-8.13} + 0.439V_2\angle{-65.77} - 0.1V_3\angle 53.23 = 0$

$-0.1V_2\angle 53.13 - 0.1V_3\angle 53.13 = 5\angle 60$

Using Cramer's rule:

$$V_3 = \frac{\begin{vmatrix} 0.439\angle{-65.77} & 10\angle{-8.13} \\ -0.1\angle 53.13 & 5\angle 60 \end{vmatrix}}{\begin{vmatrix} 0.439\angle{-65.77} & -0.1\angle 53.13 \\ -0.1\angle 53.13 & -0.1\angle 53.13 \end{vmatrix}}$$

or: $V_3 = \dfrac{2.19\angle{-5.77} + 1\angle 45}{-0.0439\angle{-12.6} - 0.01\angle 106.26} = -73.17\angle 9.57$

For the voltage source:

$$\theta_{pf} = \cos(\theta_V - \theta_I) = \cos(45 + 37.82) = 0.125$$

For the current source:

$$\theta_{pf} = \cos(\theta_V - \theta_I) = \cos(189.57 - 60) = 0.637 \text{ (absorbing power).}$$ **(Power factor)**

4. Positive reactive vars implies inductive circuit.

Reactive power:

$$Q = 120 = |\mathbf{V}|\,|\mathbf{I}| \sin(\theta_V - \theta_I) = |\mathbf{V}| * 10 * \sin(45)$$
$$\Rightarrow |\mathbf{V}| = 16.97 \Rightarrow \mathbf{V} = 16.97\angle 0\text{V}.$$
$$|S| = |\mathbf{V}| * |\mathbf{I}| = 16.97 * 10 = 169.7 \text{ VA}$$

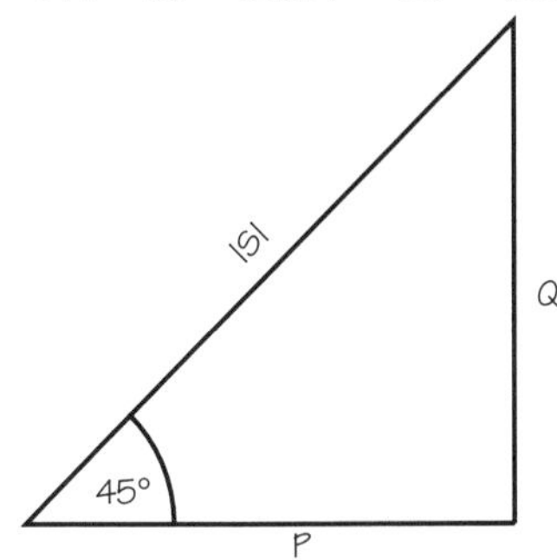

Now $P = Q \cot(45) = 120 \cot(45) = 120$ watt.

The series resistance $R = \dfrac{P}{|\mathbf{I}|^2} = \dfrac{120}{100} = 1.2\ \Omega$

And the series inductance X_L equals:

$$|\mathbf{I}|^2 X_L = Q \quad \Rightarrow \quad X_L = \frac{120}{100} = 1.2\Omega$$

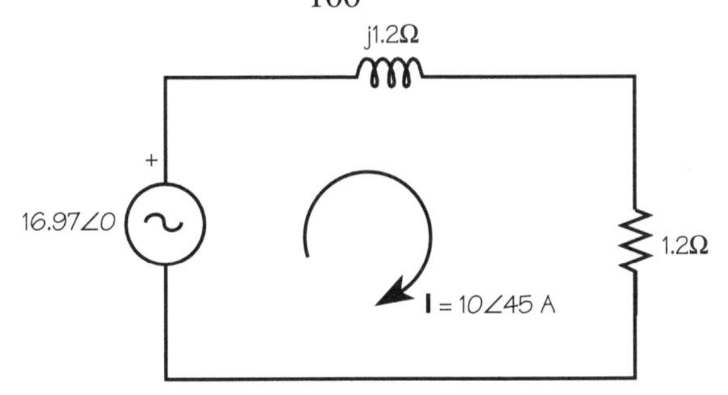

(Power triangle)

5.

$Z_1 = 3 + j4 = 5\angle 53.13$

$Z_2 = 6 - j8 = 10\angle{-53.13}$

$$Z = Z_1 \| Z_2 = 1/\left(\frac{1}{5\angle 53.13} + \frac{1}{10\angle{-53.13}}\right)$$
$$= 5.07\angle 23.96\ \Omega$$

pf = cos(23.96) = 0.913 (lag) [Inductive Circuit]

$$\mathbf{I}_1 = \frac{100}{5\angle 53.13} = 20\angle{-53.13}\text{ A}$$
$\mathbf{I}_2 = 10\angle 53.18$ A
$\mathbf{I} = \mathbf{I}_1 + \mathbf{I}_2 = 19.69\angle{-23.96}$ A
$Q_1 = |Eg|\,|\mathbf{I}_1|\sin(\theta_{Eg} - \theta_{I1}) = 100 * 20 * \sin(0 + 53.13) = 1600$ var
$Q_2 = |Eg|\,|\mathbf{I}|\sin(\theta_{Eg} - \theta_{I2}) = 100 * 10 * \sin(0 - 53.13) = -800$ var
$Q_1 = 1600$ var (Ind.)
$Q_2 = 800$ var (Cap.)

To obtain the power triangle:

$P_1 = |Eg|\,|\mathbf{I}|\cos(\theta_{Eg} - \theta_{I1}) = 1200\text{W}$
$P_2 = 600\text{W}$

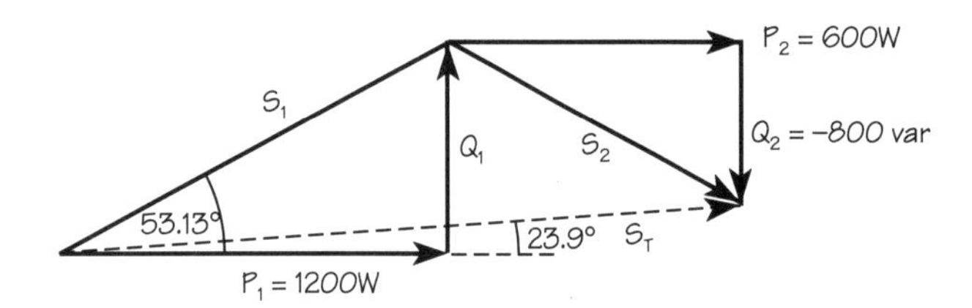

Complex power:

$S_1 = \mathbf{V} * \mathbf{I}_1^* = 100 * 20\angle 53.13 = 2000\angle 53.13$
$S_2 = 100 * 10\angle{-53.13} = 1000\angle{-53.13}$

Now, $S_T = S_1 + S_2 = 1969.77\angle 23.96$ VA
can also be obtained from:

$S_T = \mathbf{V}\,\mathbf{I}^* = 100\angle 0 * 19.69\angle 23.96 = 1969\angle 23.96$ VA **(Power factor)**

6. Drawing the Norton's equivalent circuit, we obtain (see problem 7.2. Also see problem 4.9):

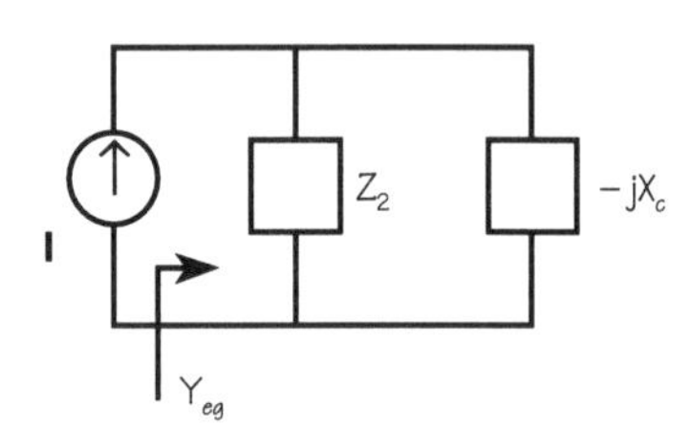

$$\mathbf{I} = \frac{14.69\angle{-42.33}}{5.428\angle 42.95} = 2.7\angle{-85.28}$$
$Z_s = 5.428\angle 42.95$
$$Y_{eq} = \frac{1}{Z_s} + j\frac{1}{X_c} = 0.184\angle{-42.95} + j\frac{1}{X_c}$$
$$Y_{eq} = 0.1348 + j\left(\frac{1}{X_c} - 0.1255\right)$$

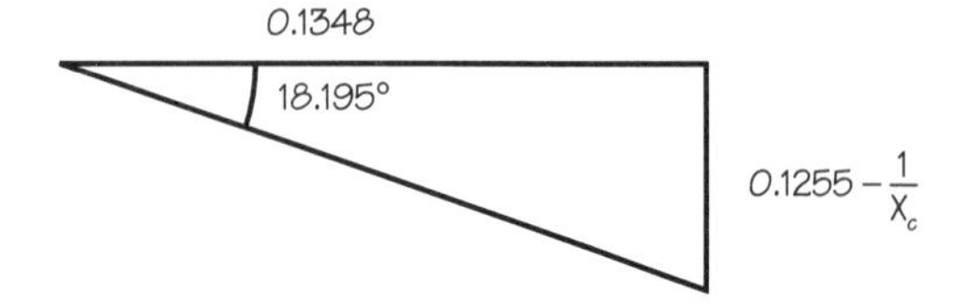

For a power factor of 0.95 (lag) $\Rightarrow \theta = \cos^{-1}0.95 = 18.195°$

$$\Rightarrow \tan 18.195 = \frac{0.1255 - (1/X_c)}{0.1348}$$

$$\Rightarrow X_c = \frac{1}{\qquad\qquad} = 12.39\Omega$$

$X_c = 12.39\Omega$ will improve the pf to 0.95 (lag).

(Power factor correction)

7. a)

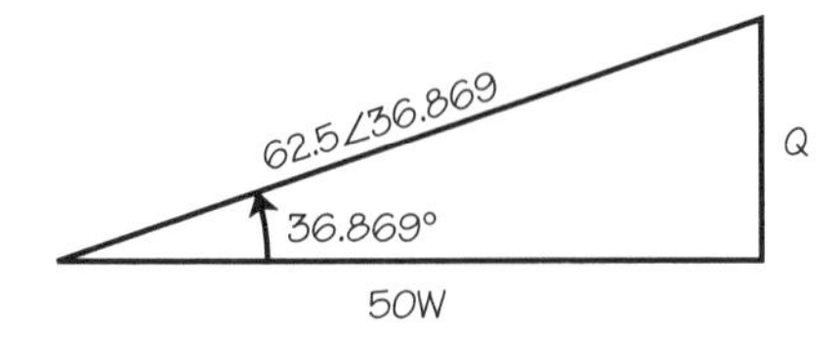

$\theta = \cos^{-1} 0.8 = 36.869$
$Q = 50 \tan 36.869 = 37.5$ vars
$S = \sqrt{P^2 + Q^2} = 62.5$ VA

b)

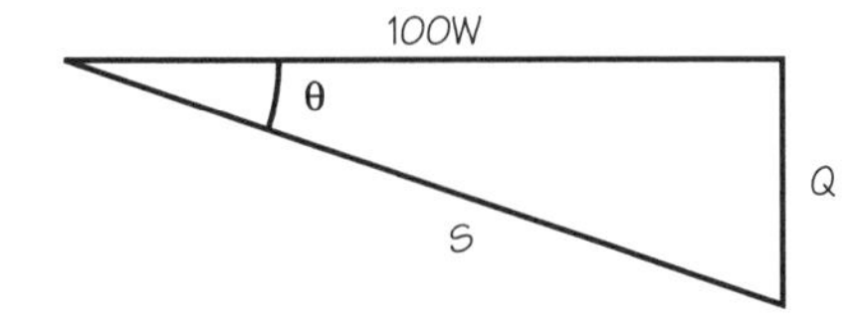

$\theta = \cos^{-1}0.9 = 25.84°$
$Q = 100 \tan 25.84 = 48.43$ vars (CAPACITIVE)
$S = 111.11\angle{-25.84}$ VA

c) $200 = \sqrt{P^2 + (100)^2} \quad \Rightarrow \quad P = 173.20W$

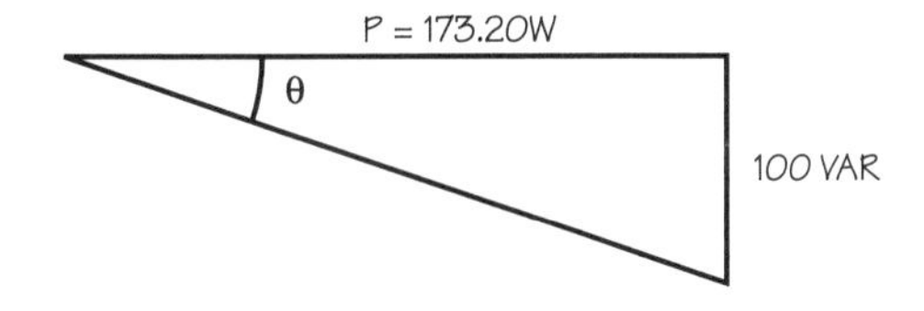

$$\theta = \tan^{-1}\frac{100}{173.2} = 30°$$

$S = 200\angle{-30}$ VA.

(Power triangle)

8. The load is characterized by 100 VA at 0.6 (lag)

$100 = 100 \bullet |\mathbf{I}| \quad \Rightarrow \quad |\mathbf{I}| = 1A$
$\theta = \cos^{-1}0.6 = 53.13°$
$Q = 100 \sin 53.13 = 80$ var
$P = 100 \cos 53.13 = 60$ watts

$$P = I^2R \quad \Rightarrow \quad R = \frac{60}{1} = 60\Omega$$

$$Q = I^2X_L \quad \Rightarrow \quad X_L = \frac{80}{1} = 80\Omega$$

$$\Rightarrow \quad Z_L = 60 + j80\Omega = 100\angle 53.13\Omega$$

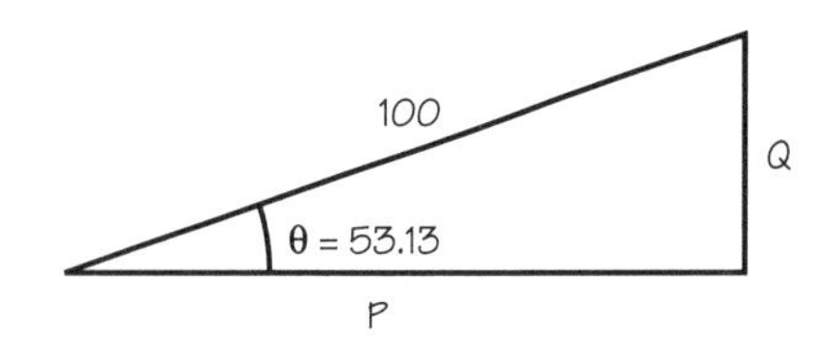

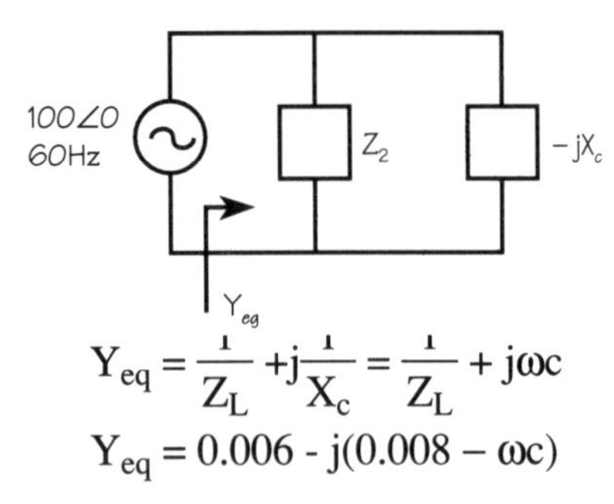

$$Y_{eq} = \frac{1}{Z_L} + j\frac{1}{X_c} = \frac{1}{Z_L} + j\omega c$$

$$Y_{eq} = 0.006 - j(0.008 - \omega c)$$

a) Final pf = 0.9 lagging

$$\theta = \cos^{-1} 0.9 = 25.84°$$

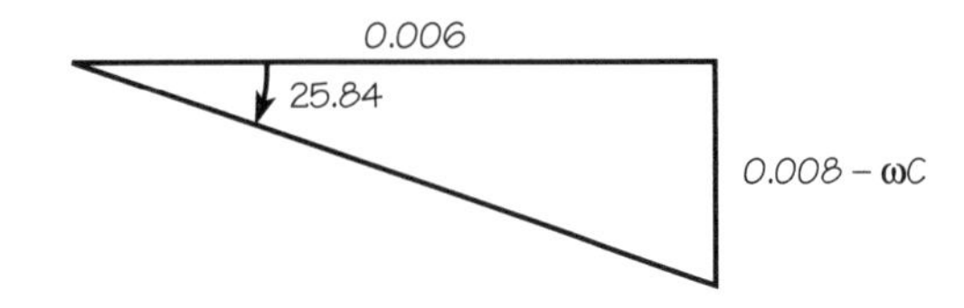

$$\tan 25.84 = \frac{0.008 - \omega C}{0.006}$$

$$\Rightarrow \quad \omega C = 0.008 - 0.006 \tan 25.84$$

$$\Rightarrow \quad \omega C = 0.00509℧$$

For $\omega = 2\pi f = 2\pi * 60$

$= 376.99$ rad/sec.

$$C = 13.5\ \mu F$$

b) $\theta = \cos^{-1} 0.8 = 36.869°$

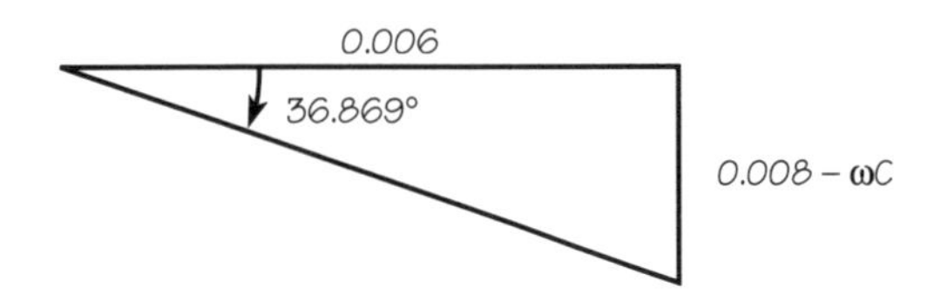

$$\Rightarrow \quad C = \frac{0.008 - 0.006 \tan 36.869}{377} = 9.28\mu F$$

c) To obtain an overall power factor of 0.8 lead

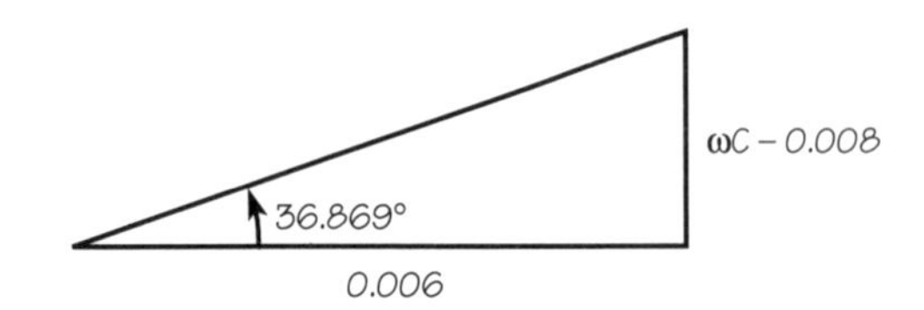

$$\omega C = 0.008 + 0.006 \tan 36.869$$

$$\Rightarrow \quad C = 33.1256\mu F$$

(Power factor correction)

Grade Yourself

Circle the numbers of the questions you missed, then fill in the total incorrect for each topic. If you answered more than three questions incorrectly, you need to focus on that topic. (If a topic has less than three questions and you had at least one wrong, we suggest you study that topic also. Read your textbook, a review book, or ask your teacher for help.)

Subject: Power and Power Factor

Topic	Question Numbers	Number Incorrect
Real and reactive power calculation	1, 2	
Power factor	3, 5	
Power triangle	4, 7	
Power factor correction	6, 8	

Spice Simulation

Brief Yourself

This chapter provides some ac and dc circuits to be solved using Spice simulation. Simulation code is provided only for a few circuits. Notations used in developing the codes are available in any Pspice manual.

Test Yourself

1. Determine the Thevenin's equivalent of the circuit shown (see problem 2.11).

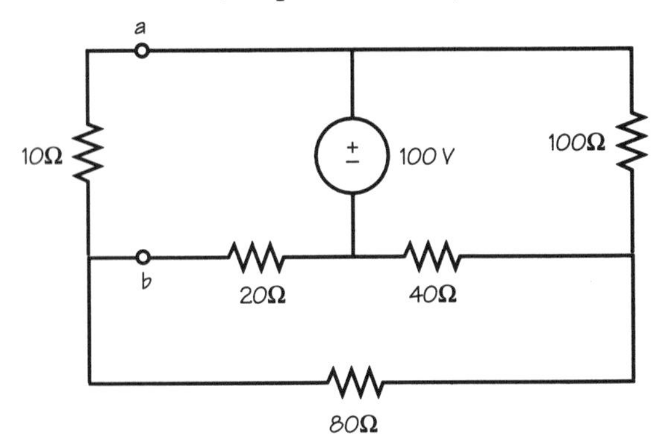

2. For the circuit shown, calculate I, using Spice. (see problem 2.15).

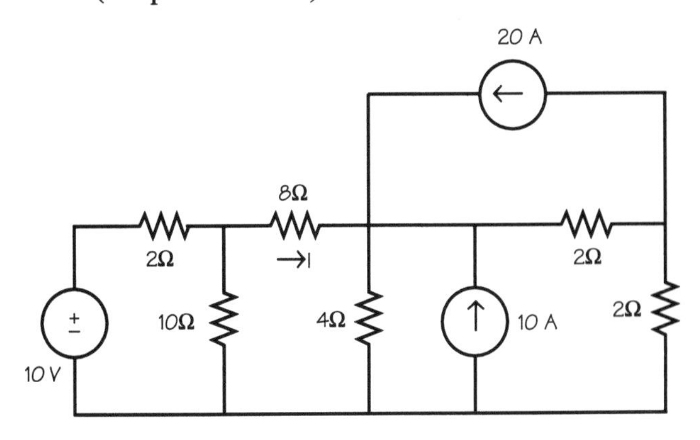

3. Replace the bridge circuit by its Thevenin's equivalent.

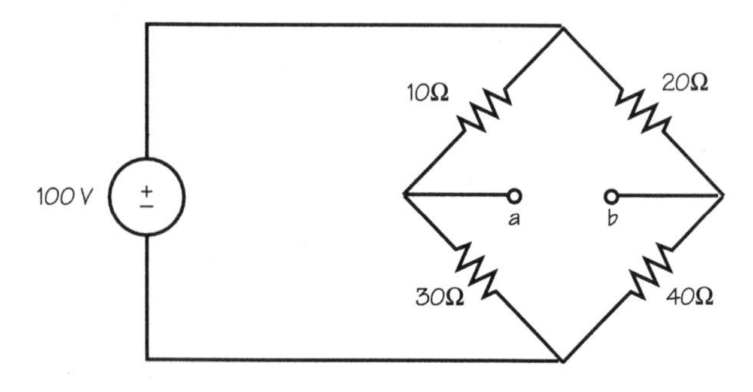

4. Using the circuit in problem 8.1, determine the load resistance to be connected between points a and b for maximum power transfer.

5. Determine $V_C(t)$

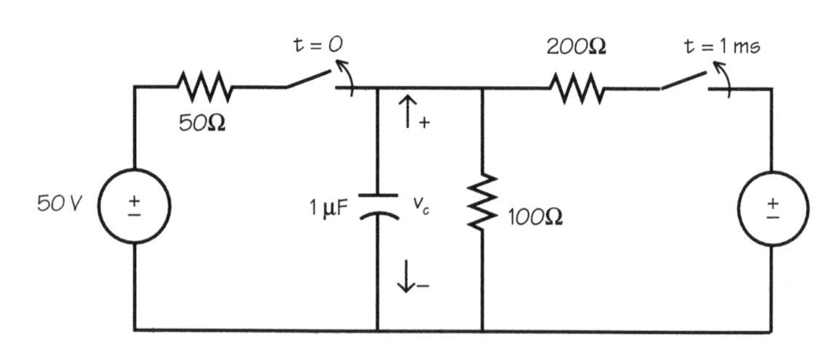

6. Calculate the voltage drop across the inductor for t > 0.

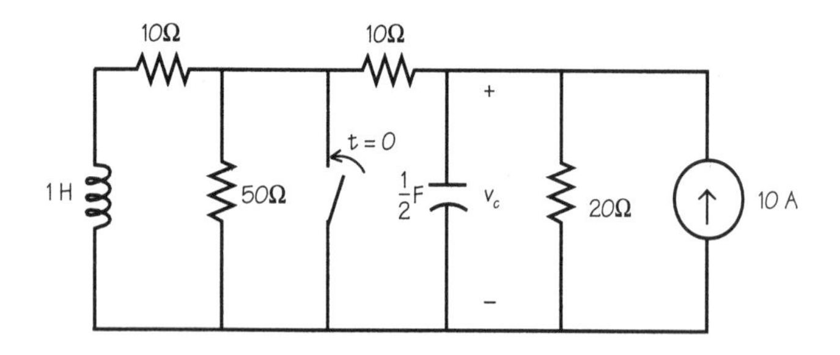

7. For the circuit shown, determine the current through the inductance.

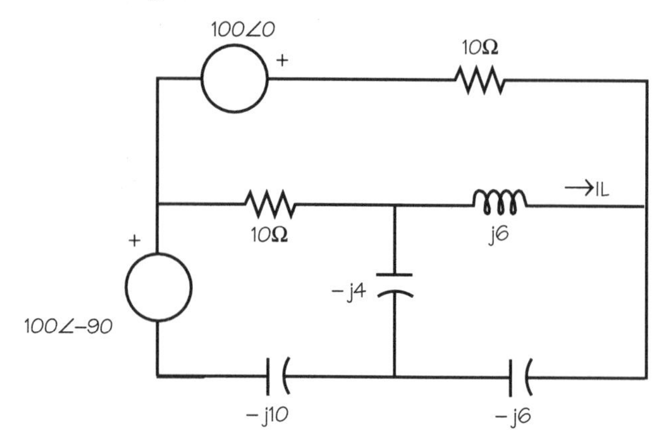

8. For the op-amp circuit shown, determine its Thevenin's equivalent.

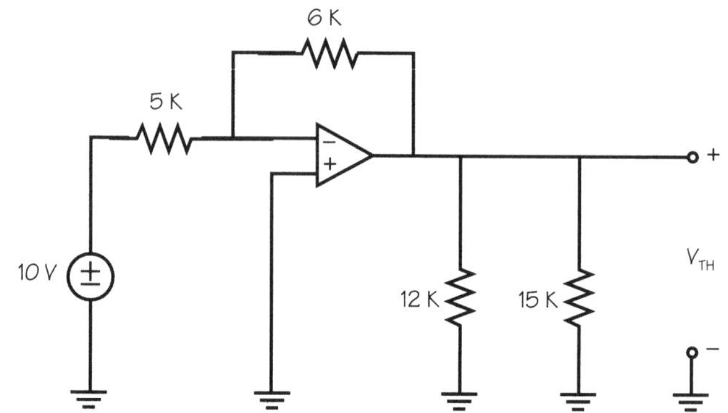

9. Determine the gain, v_o/v_{in}. Assume f = 1 kHz.

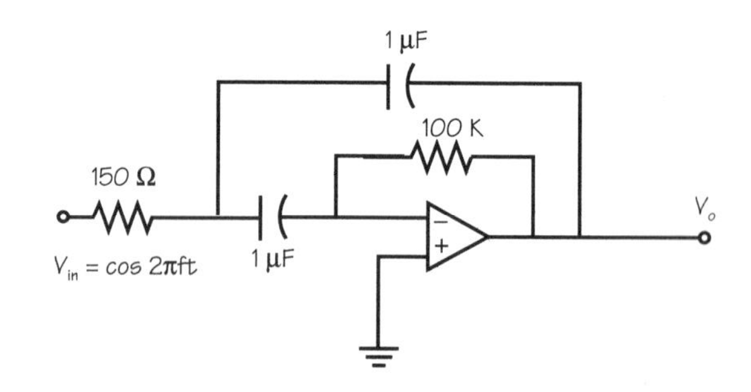

10. Determine the output voltage.

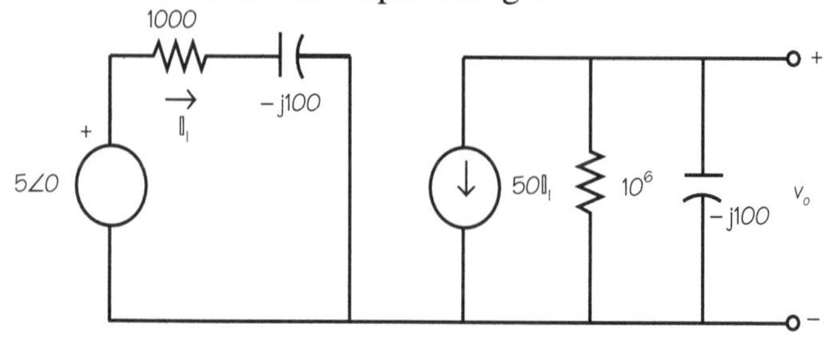

Check Yourself

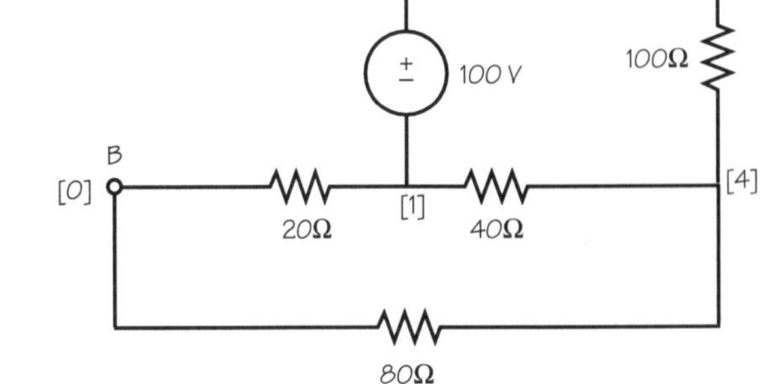

1. Thevenin Equivalent Circuit

```
v21   2   1   100
R10   1   0   20
R14   1   4   40
R24   2   4   100
R04   0   4   80
I30   3   0   0    ;Open Circuit
V23   2   3   0    ;Short Between Nodes 2 and 3
.TF V(3) V21
.END
```

Pspice Output

```
Thevenin Equivalent Circuit
v21   2   1   100
R10   1   0   20
R14   1   4   40
R24   2   4   100
R04   0   4   80
I30   3   0   0    ;Open Circuit
V23   2   3   0    ;Short Between Nodes 2 and 3
.TF V(3) V21
.END
```

```
Thevenin Equivalent Circuit
****SMALL SIGNAL BIAS SOLUTION TEMPERATURE = 27.000 DEG C
NODE VOLTAGE      NODE VOLTAGE      NODE VOLTAGE      NODE VOLTAGE
(1) -4.4444       (2) 95.5560       (3) 95.5560       (4) 17.77.80

VOLTAGE SOURCE CURRENTS
NAME              CURRENT
v21               -7.778E-01
V23               9.556E-11

TOTAL POWER DISSIPATION 7.78E+01 WATTS

****SMALL-SIGNAL CHARACTERISTICS
V(3)/v(21 = 9.556E=-01
INPUT RESISTANCE AT v21 = 1.286E+02
OUTPUT RESISTANCE AT V(3) = 1.689E+01
JOB CONCLUDED
```

(Thevenin's theorem)

2. **VOLTAGE SOURCE ON, CURRENT SOURCES OFF**

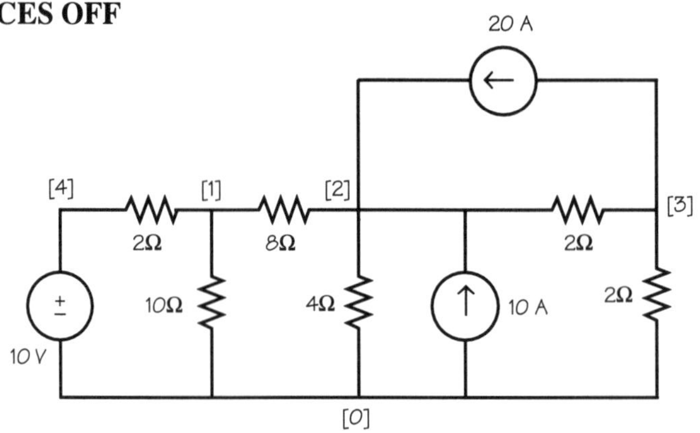


```
.OPTIONS NOPAGE
V40  4  0  10
I02  0  2  0
I32  3  2  0
R14  1  4  2
R01  1  0  10
R12  1  2  8
R20  2  0  4
R23  2  3  2
R30  3  0  2
.END
```

VOLTAGE SOURCE OFF, ONE CURRENT SOURCE ON

```
.OPTIONS NOPAGE
V40  4  0  0
I02  0  2  10
I32  3  2  0
R14  1  4  2
R01  0  1  10
R12  1  2  8
R20  2  0  4
R23  2  3  2
R30  3  0  2
.END
```

VOLTAGE SOURCE OFF, ONE CURRENT SOURCE ON

```
.OPTIONS NOPAGE
V40  4  0  0
I02  0  2  0
I32  3  2  20
R14  1  4  2
R01  0  1  10
R12  1  2  8
R20  2  0  4
R23  2  3  2
R30  3  0  2
.END
```

ALL SOURCES ON

```
.OPTIONS NOPAGE
V40  4  0  10
I02  0  2  10
I32  3  2  20
R14  1  4  2
R01  0  1  10
R12  1  2  8
R20  2  0  4
R23  2  3  2
R30  3  0  2
.END
```

Pspice Output

VOLTAGE SOURCE ON, CURRENT SOURCES OFF

```
.OPTIONS NOPAGE
V40  4  0  10
I02  0  2  0
I32  3  2  0
R14  1  4  2
R01  1  0  10
R12  1  2  8
R20  2  0  4
R23  2  3  2
R30  3  0  2
.END
```

****SMALL SIGNAL BIAS SOLUTION TEMPERATURE = 27.000 DEG C

NODE VOLTAGE	NODE VOLTAGE	NODE VOLTAGE	NODE VOLTAGE
(1) 7.1429	(2) 1.4286	(3) .7143	(4) 10.0000

VOLTAGE SOURCE CURRENTS

NAME	CURRENT
V40	-1.429E+00

TOTAL POWER DISSIPATION 1.43E+01 WATTS

JOB CONCLUDED

VOLTAGE SOURCE OFF, ONE CURRENT SOURCE ON

```
.OPTIONS NOPAGE
V40  4  0  0
I02  0  2  10
I32  3  2  0
R14  1  4  2
R01  0  1  10
R12  1  2  8
R20  2  0  4
R23  2  3  2
R30  3  0  2
.END
```

****SMALL SIGNAL BIAS SOLUTION TEMPERATURE = 27.000 DEG C

NODE VOLTAGE	NODE VOLTAGE	NODE VOLTAGE	NODE VOLTAGE
(1) 2.8571	(2) 16.5710	(3) 8.2857	(4) 0.0000

VOLTAGE SOURCE CURRENTS

NAME	CURRENT
V40	-1.429E+00

TOTAL POWER DISSIPATION 0.00E+00 WATTS

JOB CONCLUDED

VOLTAGE SOURCE OFF, ONE CURRENT SOURCE ON

```
.OPTIONS NOPAGE
V40  4  0  0
I02  0  2  0
I32  3  2  20
R14  1  4  2
R01  0  1  10
R12  1  2  8
R20  2  0  4
R23  2  3  2
R30  3  0  2
.END
```

****SMALL SIGNAL BIAS SOLUTION TEMPERATURE = 27.000 DEG C

NODE VOLTAGE	NODE VOLTAGE	NODE VOLTAGE	NODE VOLTAGE
(1) 2.8571	(2) 16.5710	(3) -11.7140	(4) 0.0000

VOLTAGE SOURCE CURRENTS

NAME	CURRENT
V40	-1.429E+00

TOTAL POWER DISSIPATION 0.00E+00 WATTS

JOB CONCLUDED

TOTAL JOB TIME 0.00

ALL SOURCES ON

```
.OPTIONS NOPAGE
V40  4  0  10
I02  0  2  10
I32  3  2  20
R14  1  4  2
R01  0  1  10
R12  1  2  8
R20  2  0  4
R23  2  3  2
R30  3  0  2
.END
```

****SMALL SIGNAL BIAS SOLUTION TEMPERATURE = 27.000 DEG C

NODE VOLTAGE	NODE VOLTAGE	NODE VOLTAGE	NODE VOLTAGE
(1) 12.8570	(2) 34.5710	(3)-2.7143	(4) 10.0000

VOLTAGE SOURCE CURRENTS

NAME	CURRENT
V40	-1.429E+00

TOTAL POWER DISSIPATION -1.43E+01 WATTS

JOB CONCLUDED

TOTAL JOB TIME 0.00

(Superposition)

3. BRIDGE CIRCUIT, THEVENIN'S EQUIVALENT CIRCUIT

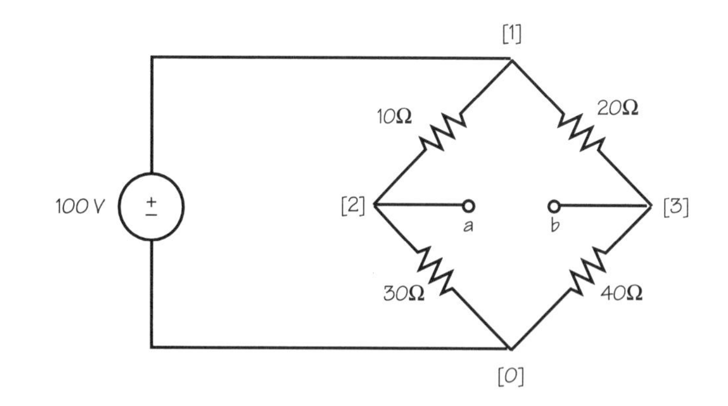


```
V10   1   0   120
R12   1   2   10
R20   2   0   30
R13   1   3   20
R30   3   0   40
I12   2   3   0      ;OPEN CIRCUIT
.TF V(I12) V10
.END
```

Pspice Output

```
NODE VOLTAGE       NODE VOLTAGE       NODE VOLTAGE       NODE VOLTAGE
(1)    120.0000    (2)    90.0000     (3)    80.0000
V(I12)/V10 = 0.000E+00
INPUT RESISTANCE AT V10 = 2.400E+01
OUTPUT RESISTANCE AT V(I12) = 1.000E+20
```

(Thevenin's theorem)

4. MAXIMUM POWER TRANSFER

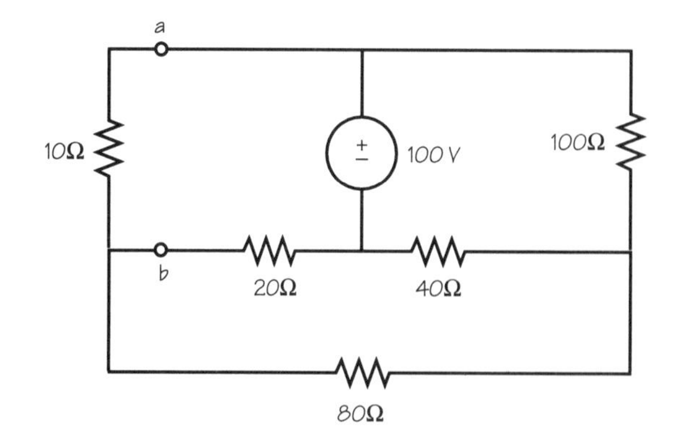


```
v21   2   1   100
R10   1   0   20
R14   1   4   40
R24   2   4   100
R04   0   4   80
RL    3   0   {R}
V23   2   3   0      ;Short Between Nodes 2 and 3
.PARAM R=8
.DC PARAM R LIST 1 4 8 12 16 20 24 30
.PRINT DC V(3) I(RL)
.PROBE
.END
```

Note: The resistance corresponding to the maximimum power is the load resistance required for max. power transfer.

(Maximum power transfer)

5. TRANSIENT ANALYSIS

```
v10   1   0   50
v30   3   0   100
r12   1   2   50
s24   2   4   a   0   swtch
vs24  a   0   pwl(0,1 lu, 0) ;upto t = 0 switch is closed
                             ;after t = 1us the switch is open
c40   4   0   1uf
r40   4   0   100
r45   4   5   200
s53   5   3   b   0   swtch
vs53  b   0   pwl(0,1 1m,0)
.model swtch vswitch
.tran 2m 2m uic
.probe
.end
```

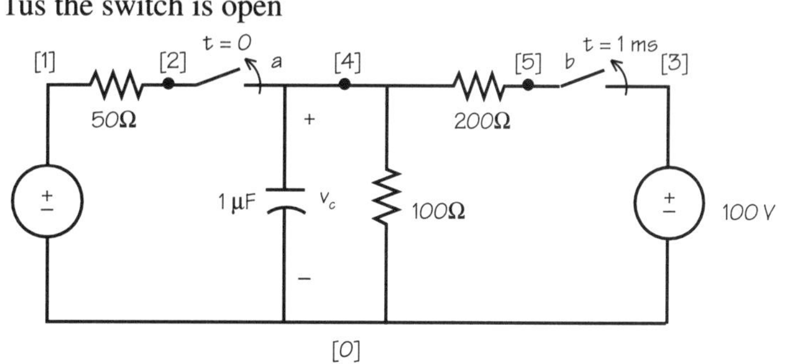

Pspice Output

```
v10   1  0  50
v30   3  0  100
r12   1  2  50
s24   2  4  a  0  swtch
vs24  a  0  pwl(0,1 1u,0)   ;upto t = 0 switch is closed
                            ;after t = 1us the switch is open
c40   4  0  1uf
r40   4  0  100
r45   4  5  200
s53   5  3  b  0  swtch
vs53  b  0  pwl(0,1 1m,0)
.model swtch vswitch
.tran 2m 2m uic
.probe
.end
```

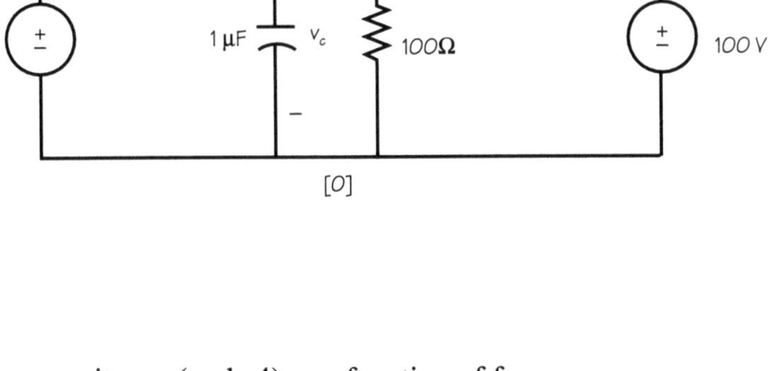

	swtch
RON	1
ROFF	1.000000E+06
VON	1
VOFF	0

Note: The graph shows the voltage across the capacitance (node 4) as a function of frequency.

(Transient)

6. TRANSIENT ANALYSIS

```
i03   0  3  10
l10   1  0  1H
r12   1  2  10
r20   2  0  50
r23   2  3  10
c30   3  0  0.5f
r30   3  0  20
s1    3  0  a  0  swtch
vs1   a  0  pwl(0,1 1u,0)
.model swtch vswitch
.tran 10m 10m uic
.probe
.end
```

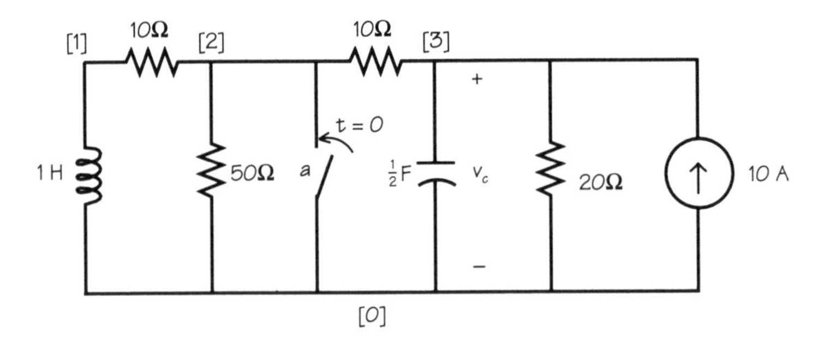

Pspice Output

```
i03   0  3  10
l10   1  0  1H
r12   1  2  10
r20   2  0  50
r23   2  3  10
c30   3  0  0.5f
r30   3  0  20
s1    3  0  a  0  swtch
vs1   a  0  pwl(0,1 1u,0)
.model swtch vswitch
.tran 10m 10m uic
.probe
.end
```

```
         swtch
RON      1
ROFF     1.000000E+06
VON      1
VOFF     0
```

(Transient)

7.

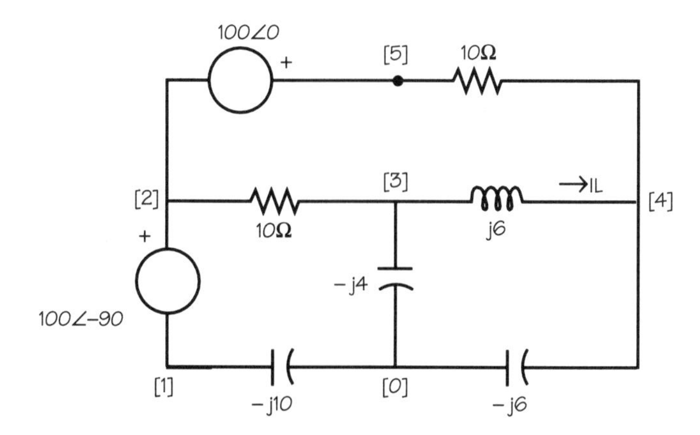

(Phasors)

8.

```
.subckt op_amp n p output
rin n p 10meg
rout c output 100
egain c 0 p n le5
.ends
vs    1  0  dc  10
r12   1  2  5k
r23   2  3  6k
r30   3  0  12k
r301  3  0  15k
i30   3  0  0
xoal  2  0  3   op_amp
.tf v(3) vs
.end
```

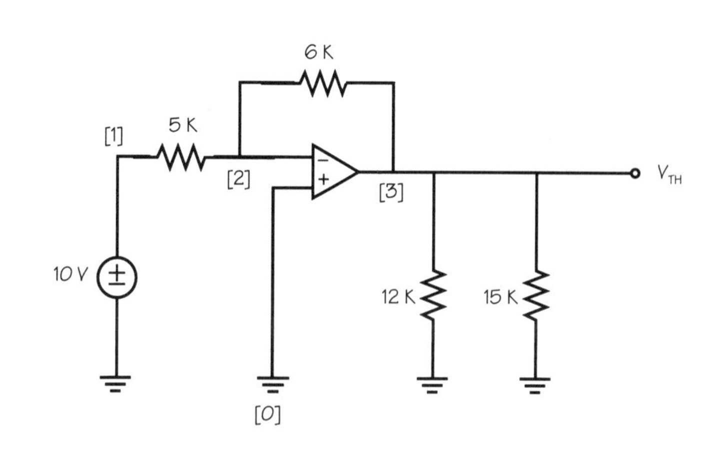

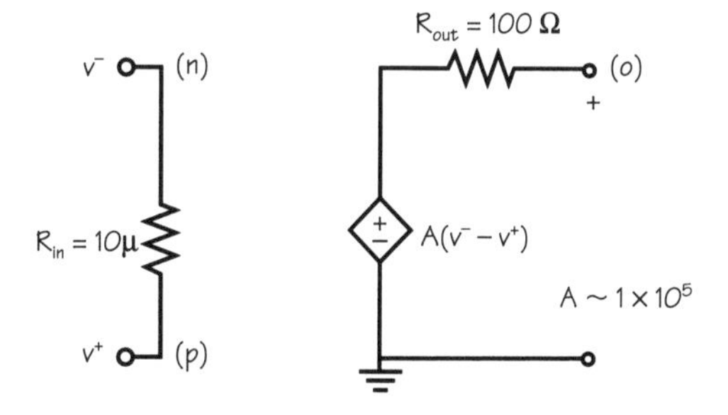

Pspice Output

NODE VOLTAGE	NODE VOLTAGE	NODE VOLTAGE	NODE VOLTAGE
(1) 10.0000	(2) 123.8E-06	(3) -12.0000	(xoal.c) -12.3800

VOLTAGE SOURCE CURRENTS

NAME	CURRENT
vs	-2.000E-03

TOTAL POWER DISSIPATION 2.00E-02 WATTS
****SMALL-SIGNAL CHARACTERISTICS
V(3)/vs = -1.200E+00
INPUT RESISTANCE AT vs = 5.000E+03
OUTPUT RESISTANCE ATR V(3) = 2.201E-03

(Operational-amplified circuits)

9.

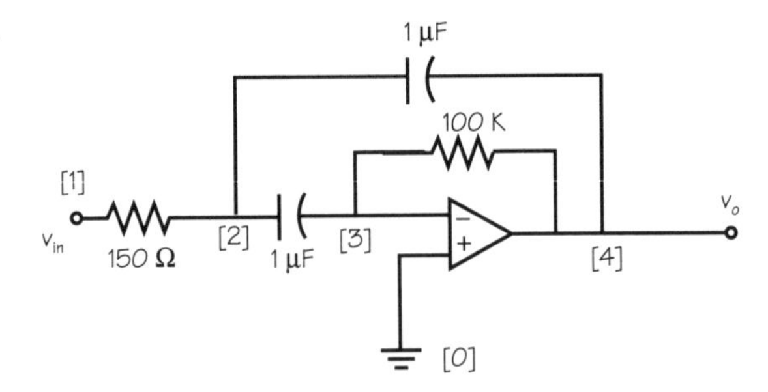

(Operational-amplified circuits)

10.

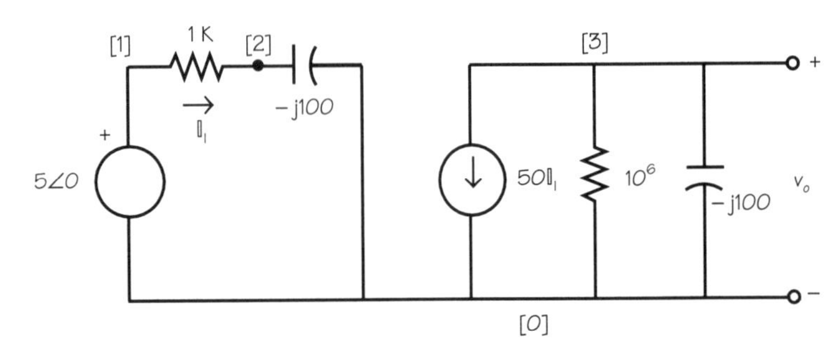

(BJT circuit)

Grade Yourself

Circle the numbers of the questions you missed, then fill in the total incorrect for each topic. If you answered more than three questions incorrectly, you need to focus on that topic. (If a topic has less than three questions and you had at least one wrong, we suggest you study that topic also. Read your textbook, a review book, or ask your teacher for help.)

Subject: Spice Simulation

Topic	Question Numbers	Number Incorrect
Thevenin's Theorem	1, 3	
Superposition	2	
Maximum power transfer	4	
Transient	5, 6	
Phasors	7	
Operational-amplifier circuits	8, 9	
BJT circuit	10	